"TAPE'S ROLLING, TAKE ONE"

The recording life of

Adrian Kerridge

*Six Decades of Recording and Producing,
from the Rock 'n' Roll Years to
TV Scores & Blockbuster Movies!*

Volume One – 1940s, '50s and '60s

M-Y BOOKS PAPERBACK

© Copyright 2016
Adrian Kerridge

A CIP catalogue record for this title is
available from the British Library

ISBN–978-1-911124-22-1 (POD)
978-1-911124-23-8 (Ebook)

Dedication

This book is dedicated to my wife Mary for her love, patience, understanding, and encouragement when I spent so many hours in the studio, recording and producing music in England and Europe; also to my children Kathryn, Suzanna, Virginia and Nicholas.

I would also like to dedicate this book to my dear friend and business partner Johnny Pearson, who sadly passed away in March 2011 after a brief illness, and with whom I enjoyed so many memorable and creative times in the studio. A very successful pianist, composer and arranger, Johnny was a man of integrity and humility, and a wonderful human being.

In addition, I would like to acknowledge the part played by Peter Gregory – group financial director, friend and colleague of some forty years' standing. A brilliant and thoroughly creative accountant, Peter was of enormous help and kept a close watch over the expenditure, taking much financial pressure off my back. His skills proved an ideal complement to Johnny's musical talents, and he was always there when needed, without question. He passed away suddenly in October 2007 aged 59.

Madeira November 2013

Foreword

By Dave Clark

People talk about the late, great Sir George Martin as being The Fifth Beatle and he certainly deserves such an accolade, not least for producing their legendary and ground-breaking records. By that same token, Adrian Kerridge surely has to be the sixth member of The Dave Clark Five (The DC5). While still in his twenties, Adrian totally got what we were all about and thanks to his empathy for The DC5 as artists - along with his technical wizardry in the studio - he managed to capture the essence of our live stage sound on analogue tape.

Who could have known back in those heady days of the 1960's when we were all so young that fifty years on our records - with Adrian at the studio controls - would stand the test of time and still manage to thrill people?

Bruce Springsteen has publically stated that he is not only a DC5 fan but that he studied some of the studio techniques Adrian and The DC5 developed and applied them to some of his own recordings. Praise comes no higher! Indeed in a filmed interview Bruce generously said, "The DC5 made some of the greatest Rock And Roll records ever made; those were big, powerful, nasty-sounding records, man; a much bigger sound than, say, The Stones or The Beatles. They were thrilling, inspiring, simply exciting. To this day they are still great productions."

In the studio Adrian was a one-man band with no assistants; not only was he the engineer and tape operator but he also set up, maintained and fixed all the studio equipment as well as the

DC5's equipment. When in 1998 the Association of Professional Recording Services honoured Adrian with their prestigious Lifetime Achievement Award - as they had done with Sir George Martin the previous year - it gave me great pleasure to present the award to Adrian. That recognition from his peers for his ground-breaking technical achievements confirmed what The DC5 and their fans worldwide had always known from the evidence of our own ears: that behind the scenes of Rock And Roll's great era, Adrian was a true creative innovator.

Contents

Introduction

Adrian Kerridge, recording engineer, producer and studio owner, presents his personal view and history of the music recording industry, tracing his long career from its analogue tape beginnings in the 1950s, through to the digital hard-drive recording of major film scores in the 21st century. Unique in its scope, it covers the bands and artists, engineers and personalities, techniques and technology.

The story encompasses Adrian's work with the legendary Joe Meek, his recording of the UK bands that spearheaded the 1960s invasion of the US charts and his involvement with mixing console design.

A long and successful career:

Awarded some 300 Platinum, Gold and Silver discs across multiple musical genres, with recordings of the Dave Clark Five alone resulting in worldwide record sales of 100 Million. (Source: Dave Clark)

Recipient of Lifetime Achievement Award from the Association of Professional Recording Services (APRS), in recognition of a lifetime's service to the music industry. One of only two such awards – the other being awarded to the late Sir George Martin.

Gold-badge Merit award from British Academy of Songwriter & Composers, for services to the music industry.

Fellow Institute of Professional Sound (Broadcast)

1

Fellow Association Professional Recording Services (Studio Recording)

Life Member Audio Engineering Society.

Member Association of Motion Picture Sound. (Music scoring).

Retired previous member British Kinematograph, Sound and Television Society serving the technical and craft skills of the film, sound and television industry.

Retired previous member Institute of Directors.

Retired previous member Institute of Management.

Chapter 1
Growing up with Music

The month of March 1938 was unusually dry and sunny for the time of year, with an average temperature of 8°C. At Harrow-on-the-Hill in North West London, in a private nursing home at 100 High Street, a boy was born to Margery and Leslie Kerridge. Christened Adrian Nicholas, their only child, he arrived eighteen months before the outbreak of the Second World War.

Switches and Bombs

My family lived in Northolt, Middlesex, in a new semi-detached house at no. 7 Harewood Avenue, bought by my father Leslie for £675 (around £48,000 today) in 1935 shortly after my parents married at St Mary's church Northolt Village. There was no central heating instead we had a fireplace in every room and we had mains electricity. Coal was stored in a coal bunker outside containing Anthracite for the boiler and coal for the fires. Some homes in the surrounding area had no electricity and relied on gas lighting! Those with no electricity and to listen to the wireless they had battery models which had lead acid accumulators, with glass sides, for the valve heaters and high tension batteries for the valve plates. To recharge the accumulators they had to take it to the local radio store. We had a telephone – with the number WAX 1927; not everyone had a telephone in those days. We had that privilege because my father was in a protected occupation during World War Two maintaining and repairing telecommunications in London that had been put out of action by the Nazi bombing during the battle of Britain. Prior to the start of WWII, my father was employed by Reliance Telephones to service PBX (private branch exchange) systems which also connected to the UK General Post Office network. These were to be found in large companies, and had one switchboard operator or more, depending on the size of the firm, to route calls internally and manually. By today's standards these were archaic and needed much maintenance, but they evolved into the computerised PABX (private automatic branch exchange) digital systems that are now all too common, using the somewhat notorious automated messages and menus – push this button, then that, and finally one gets to speak to a human! My father, before outbreak of war, maintained Selfridges' PABX in the basement of the store.

At war's end, my father changed jobs and travelled around Southern England maintaining and repairing traffic light electronics, which were housed in cabinets that were usually located on windy street corners. Sometimes I would travel with him and experience his working outside in all weathers – not pleasant. The traffic signals in those days were operated by electro-mechanical devices – relays. He subsequently changed jobs back to telephone engineer, maintaining and repairing broken equipment in telephone exchanges, which had a vast array of electro-mechanical switchgear invented in 1888 by an American, Almon Strowger. He patented the first automatic exchange in 1891. Born in 1839, he was an undertaker in Kansas City, Missouri! The Strowger switchgear (known as uniselectors), of which there were many, chuntered away with incredible noise that could be heard outside the exchange walls. They needed much cleaning of their multiple switch contacts, which were arranged in 10 levels, otherwise there was a possibility of a misrouted call or no connection at all. Each level had 10 contacts arranged in a semicircle, with these devices connecting one caller to another via impulses from the rotary telephone dial – a slow and cumbersome method. Even aged ten, with my first experience of electro-mechanical technology devices, they held a special fascination for me. There were huge lead-acid batteries, connected in series to provide a nominal 50 volts DC to power the exchange in the event of a mains failure.

The exchange was in Shell-Mex and BP Ltd Company House, to provide for the internal telephone system located in the sub-basement. It was quite a noisy place with all the Strowger electro-mechanical switches operating. The ringing tones were generated by two mechanical rotating machines driven by small electric motors with a shaft that engaged the contacts. These operated in sequence, to provide the familiar *ring-ring* sound of old analogue telephones – difficult to imagine in the today's digital age!

We lived one mile from Northolt airport where, at the outbreak of war in September 1939, were stationed a squadron of Polish airman who flew Hurricane aircraft, and what a good job the fearless Poles did in dogfights attacking the Nazi aeroplanes. During the war, I was woken most nights by the Luftwaffe trying to attack Northolt aerodrome. Every morning at about 5.00, I was awoken yet again by the Hurricanes' engines warming up, ready to go at a moment's notice during the height of the Blitz.

After the war, in May 1955, a memorial was erected to those Poles who flew out of RAF Northolt: "The Polish War Memorial", on the left hand side of the A40 adjacent to the slip road to Ruislip.

My father worked in his normal job by day and on many nights was an ARP (Air Raid Patrol) Warden, patrolling the streets to ensure blackout was maintained. He wore the regulation tin hat and would often bring home shell shrapnel; heavy, evil-looking stuff. While on patrol during a heavy raid, he would briefly return to the house and ask my mother through the letterbox, "Margery, are you and the boy alright?" During a raid my mother and I would take shelter under the stairs. It was deemed the safest place to be if there was a direct hit on the house, and living near RAF Northolt that was a distinct possibility. There was a concrete built air raid shelter very near our house – it was damp and it stank so my mother refused to go there. My mother was a seamstress by trade who gave up her job when I was born and spent many hours behind her electric Singer sewing machine at home, modified from treadle power, making dresses to supplement the family income.

When I was very young, and heard any music played by an orchestra on the wireless (as it was then called), I imagined the band was in the radiogram's loudspeaker! I listened every night to *Children's Hour*, broadcast on the BBC Home Service from 5 to 6pm with "Uncle Mac" (Derek McCulloch). *Children's Hour* was one of my favourites, especially the "Larry the Lamb" character. When other regions joined the London broadcasts, the engineers were warned to be aware of sound levels. I guess this was because there was no transmitter protection – no limiter in the signal chain – and any over-modulation would cause the transmitter to come off air. Perish the thought for dear old Auntie Beeb! Derek McCulloch signed off air every night with the line "Goodnight children everywhere".

Despite the strife of the war years, and the shortages that went with them, I had a very happy childhood, despite the deprivations of war, with caring parents always working and a few toys built by my father, my favourite, a sit-on wooden model of a road steam roller. Those early years got progressively harder for all, thanks to strict food rationing imposed in January 1940. Foods considered a basic now such as oranges or bananas were scarce and rarely seen, let alone any of the exotic fruits that grace the shelves of today's supermarkets. Food rationing was strictly enforced by the ration book, one per person with coupons that were exchangeable for foodstuffs, such as meat (except sausages, which were hard to come by), eggs – one fresh egg a week if available (unless you were lucky enough to keep chickens, in which case eggs were plentiful!) – milk, butter, and cheese. If you were a farmer there was not as much of a problem with rationing as there was in towns and cities (unless you went to the black market) because they had the animals to kill and eat – pigs, cows, sheep – and made butter. Rabbits were plentiful though! We had so many to eat I got fed up with eating them. Even today, when I see rabbit on a menu it's a no-no for me! Even clothes were rationed, known as standard clothing, and were marked by a CC41 label and young children were given small bottles of concentrated orange juice as a vitamin C supplement. It wasn't until sometime after the war's end that I saw an orange; I never knew what a banana was or even saw one until they returned to the greengrocers' shops – hard to believe now. My father grew as many vegetables as he could in the front and back gardens – the cabbages sometimes decimated by caterpillars. The whole population was encouraged to do this, using every available piece of land, inspired by the "Dig for Victory" slogan. Those circumstances are difficult to comprehend now in an age of immediacy and plenty. I clearly recall, during the height of the Battle of Britain, playing with my friend Pamela Yost from across the road. She had a brother, Peter, with whom I was friendly, but I found it quite odd that his father and mother always addressed him as *Boy*: "Come here *boy* do this and do that *boy*." I only saw their father once and had no idea what he did for a job.

During the Blitz in summer 1941, while playing in my back garden, Pamela and I would often watch the dogfights going on very high in the sky with the swooping and diving planes (only after the war did I learn the planes were Spitfires and Hurricanes) painting vapour-trail patterns across the clear blue sky. Since we lived so close to from RAF Northolt my back garden gave us a ringside seat. It was remote and surreal but, as children, we were not as frightened as during the night-time bombing raids; never knowing when the next raid would come and what devastation it would bring about. Reflecting back the night time air-raid-sirens put a chill down my spine. We would wait anxiously for the relief of the all clear siren: no attack on RAF Northolt aerodrome but attacks elsewhere.

My father was a keen amateur photographer – he won some awards at the local camera club for his black and white photography taken around the streets of London pre war – and a member of a church choir in Roxeth Church, Harrow-on-the-Hill. He also learned to play the piano and studied music theory and composition, for classical and church music. We had a very extensive collection of twelve inch 78s – Mozart, Brahms, Beethoven and so on. Being 12″ 78s, and some of the works being quite long, they were in boxed sets of four or five records. So, for example, on reaching the end of the first side the second side started as an edit, and so on with all the sides to the end of the work! It always fascinated me that when the records were played the needle was not steel but a natural thorn, which could be sharpened by a small round hand tool. It was explained to me that when you used steel needles too many times on shellac records they could damage them by cutting into the grooves, so that gramophone record took on a shade of grey, which degraded the fidelity. I spent many hours with my father listening to classical works. My favourite was the last movement Beethoven's Symphony No 4 in B flat Major Should be Symphony No 6 in F Major (Pastoral) – a good tune! Herr Hitler would have approved, accordingly the three master composers that represented good German music in his view were Ludwig van Beethoven, Richard Wagner and Anton Bruckner. Nonetheless, Hitler particularly favoured Wagner, who was uncompromisingly anti-Semitic and loved by the German people and it has to be remembered that these three composers lived prior to the 20th century. However, a point of interest, *Beethoven's Symphony No. 5 in C minor's* opening began with a four-note, short-short-short-long intro, which repeated twice within its five-bar length:

Those of you who studied or remember WWII history, will be aware that Churchill used the V sign with the palm of his hand pointing outward (not the V sign some use today): it signified V for Victory. V in Morse code is "dot-dot-dot-dash" sounding "da-da-da-dumm, da-da-da-dumm" – the four notes over five bars. During the war, the BBC adopted those four notes in its broadcasts to Europe they also transmitted messages personnel after the evening news: coded messages ("Yvette likes big carrots …Paul has some good tobacco … the secretary is very pretty").... Mademoiselle strokes her dog's nose: source BBC) to the resistance and others in occupied France to alert the listener to a parachute drop of weapons or an agent where the resistance could locate them. The BBC recording was played by percussionist Jimmy Blades (I later met him on sessions when I was at IBC) on a damped African membrane drum with a tympani stick, to signify V for victory. (Source: Graham Melville-Mason, *Independent*, on James Blades). The short notes were damped by his hand on the drum skin. Blades was often at IBC as a session player on percussion. Any persons in Europe painting the V sign in public places, if caught in the act, would be severely punished, and even more so in the occupied UK channel islands. How ironic that the music of one of Hitler's favourite composers was used against him!

When just five years old, I went to a local infant school within easy walking distance of my home, and recall we had chalk and small slate boards to learn to write on. It was small class of not more than ten infants and one teacher. It was a relief when "Victory in Europe" (VE day), war's end, was announced on Tuesday 8th May 1945. Up to then, as a young child I had only known a state of war, fuelled by the constant attacks on the country. Part of the relief was the realisation that my father no longer had to spend his nights patrolling the local streets, to ensure no light was shining from people's homes to the street thus denying the Luftwaffe any ground reference points to establish their aircrafts position. At last he would be with us at home and not outside in constant danger from bombs and incendiary devices. During one raid an incendiary (phosphorus) bomb no doubt intended for Northolt Aerodrome landed on the wooden roof of our coal store. Fortunately, on this occasion, my father was at home and it was soon extinguished by throwing buckets of sand over the fiercely burning device. The end of the war meant no more Nazi enemy planes attacking Northolt aerodrome, no more big guns booming in the middle of the night (there were two large anti-aircraft guns not more than a quarter of a mile from my home) to defend RAF Northolt and, best of all, no alarming whine of air raid sirens. As a child, the thought of what the next raid would bring was quite terrifying. I remember one night being in my mother's arms in the kitchen when a raid was on – an attack on the aerodrome. One enemy plane was so low that I remember seeing the pilot very clearly; he dropped a bomb that hit a house up our road on the hill overlooking the aerodrome – a lousy shot! For a few years after the war, whenever I heard recordings of the air raid sirens on the wireless it sent a left me cold.

I moved to Northolt primary school in Spring Term 1946, again within walking distance of my home. The teachers were brilliant. We had gas lighting in the classrooms, which struck me as very odd, as we had electricity at home. In the Junior part of the school there were separate hut-type buildings heated by old fashioned coke stoves in each class, which when stoked up glowed red on the outside – no fire guards around them. ('Elf 'n Safety would have a fit today; those box tickers would love it!). The government of the day must have been promoting milk for children's health and well being because one day a camera unit arrived to film us drinking it – my first encounter with a film unit. Third-of-a-pint bottles were issued free every day, and along with another boy I got the job of delivering them to every classroom on a four-wheel handcart; the attraction being we could whiz around the playground on it, except during the terrible winter of 1947 when there was very deep snow! One clear memory I have is walking to school in deep snow then through the entrance to school playground, accompanied by my mother, through snow piled up either side of a cleared pathway at least six feet in height on either side!

Looking back, I felt such a sense of liberation when the war in Europe ended on 8th May 1945. I could play in the street once again, and walk the fields in spring and summer with my friends and enjoy the bird song – even they sounded free! We looked for birds' nests in the hedgerows but didn't take the eggs: instead we just tried to identify the species – blackbird, sparrow, thrush and so on. We climbed trees and enjoyed the fresh smells of countryside, freshly cut hay and the reaping of corn, the dry dust rising and rabbits, hares and small field mice running out for their lives from the advancing tractor. Harvest time saw us watching the cutting and threshing of the corn, the thresher driven by a steam tractor. We climbed up the straw ricks (no straw bailing in those days!) and generally enjoyed ourselves, much to the dismay of the farmer – an ill-tempered rotund man, who took exception to our sliding down the ricks and chased us away red-faced and shouting. What very happy days they were – to be free after war.

The Harvest service in church on an autumn Sunday was something we boys looked forward to, the church decorated with fruit, vegetables, flowers and wheat sheaves as well as the Harvest Supper at the church hall, usually on the previous Saturday. I used to go to Northolt village, as it was then, play in the stream close to the thatch-roofed pub, where the beer was delivered in oak barrels by a steam dray. Behind the village church was an old hollow oak tree we used to climb inside and play inside and around it, a large pond where we went newting (what beautiful creatures!). My mother said, "Don't bring them home". So we didn't!

My father loved to fish and, on occasions, we would take the No. 140 bus to Hayes railway station to catch the Great Western Railway (GWR) steam train from Hayes to Denham accompanied by the smell of coal smoke, oil and emitted exhaust steam and if you looked out of the window, which I did frequently, the occasional black smuts in the eyes from the engine. We would walk through the fields to the small crystal-clear river Colne running alongside the Grand Union Canal. We caught trout, usually three or four, took them back home and my mother would cook them for our tea. I've loved trout ever since – especially the smoked variety.

Summer also saw the fair come to the local Northolt Park, and one particular attraction was the huge steam-driven showmen's engines towing the fairground rides and the showmen's caravan living quarters. I was fascinated by the large dynamos mounted on the front of the engines, driven by wide belts from the flywheel with the large round volt and current meters, located near the dynamo on the front, reading a DC voltage of 115/125 volts and the current drawn, which varied according to the type of fairground attraction it was driving. For the dodgems, it was about 50 amps, carried by great big thick cables – so the load on the dynamo meant the engines had to work really hard, chuffing heavily.

I began to learn piano at seven but soon got fed up with playing endless five-finger exercises and learnt to recognise the notes by hearing them – my father would ask me which was this and that note, which C, which was a sharp or a flat, et cetera, without looking at the piano keyboard. It was all in good-humoured spirit as it should have been – and it was good training for the ear. I also learned to read music – or at least understand some of scores that my father collected, such as Handle's oratorio *Messiah*, which I sang in the church choir when I was older. My favourite was the *Alleluia* chorus, in parts not easy to sing because its long passages without room for taking a breath made good breath control essential. To work around the breath problem, the choirmaster would decide who would take a breath at what bar in the score. This meant that long passages could be sung without large gasps, and the piece sounded seamless. And when you took a breath, it had to be a "quiet" one! In my later years this was a lesson learnt when directing vocals or vocal groups using close-mic techniques. While I was still seven years old, I joined my father's church choir as a junior choirboy – by that time I was attending Northolt primary school. The choirboys had singing practice on Wednesday, and the full choir on Friday evenings, and although our ferocious choirmaster took no prisoners, it was excellent training: "Learn to project the voice, breath correctly, don't sing from the throat, or move the voice on certain notes from stomach to throat, sing from the stomach and feel the diaphragm moving, and oh, don't forget to pronounce the ending of words clearly and ah, as a choir, no machine gun endings please, all the Ts, Ds and Ss should end together. No mumbling of the sung word. Remember your diction! And yes, one other point on the unaccompanied pieces, try not to sing flat!"

Little did I know then that all that good advice would stand me in good stead for the future, when dealing with artistes in the studio, some of whom thought they knew it all!

Unfortunately for the Church Verger we choirboys were no cherubs. The Verger was a portly gentleman who lived in a bungalow at the entrance to the church drive – we called him Foxy. As part of his duties he was responsible for the upkeep of the church grounds and graveyard. He hated us boys trampling over his freshly cut lawns and put iron hoops in the ground to deter us. Every Wednesday and Friday, in the grass-growing season, we would pull out the hoops and scatter them on the ground, leaving holes in the lawn, much to his annoyance. He complained to the vicar and we were warned but "boys being boys"... On summer Sundays whilst waiting for evensong

13

to start we would often sit outside the vestry on Foxy's grass and play five stones – an old game played by children at that time. To play we selected five small round stones and whoever's turn it is throws them into the air and tries to catch as many as possible on the back of the same hand. The *stones* that were caught are thrown up again from the back of the hand where they came to rest and as many as possible are caught in the palm of the same hand until the player fails to catch any and is "out".

On some occasions after choir practice in late autumn/early winter, when household coal fires were lit, we had to walk home to Northolt because the thick smog that enveloped the area brought the buses to a halt. It stuck in the throat and there was a strong taste of sulphur in the mouth – there were many coal-burning fires in those days and, when the weather conditions were right, the smog appeared frequently a thick yellow in colour, causing one to cover the mouth and nose.

During my early choir period I joined the Cubs – but I didn't last long. My simply being too boisterous is one reason that comes to mind – and I didn't take it seriously. All that *dib dib* stuff! I was asked to leave and didn't really mind because as far as I was concerned I had better things to do with my time. The best part was buying fourpenn'orth of chips after Cubs from the chippy (fish and chip shop) across the road. To this day I still love the smell of fish and chips with vinegar and salt – and it's got to be eaten out of a newspaper!

Something new for me after the war was going on holiday. During the war, it was impossible to go on holiday, as many of the beaches on the South and East coasts were strictly off limits. They were located in public exclusion zones, and some were mined because in the early war years a German invasion was a distinct possibility.

My first holiday, of the two spent on the Isle of Wight in 1947 and 1948, was with my mum and dad in Ventnor. We travelled by Southern Region main-line train to Portsmouth, taking the steam-driven paddle steamer to Ryde Pier head – a journey across the Solent of about half an hour – and then catching a numbered and named (which fascinated me: No.16 Ventnor, No. 21 Sandown and so forth) steam train to Ventnor. As a child of nine years, having not been that far afield before, I was excited to be seeing another place – and by having the freedom to roam by the seaside! I clearly remember as the train slowly made its way across the island, looking out of the right hand side of the carriage I saw a row of three strange-looking high masts on a hill in the distance. Naturally, having an enquiring mind, I thought Dad would know what they were for; he told me they were something to do with radio during the war. It wasn't until many years later, still being inquisitive, that I discovered they were part of what we now know as radar (**RA**dio **D**etection **A**nd **R**anging). It was then known as "home chain radar", later supplemented by "chain home low" to detect aircraft as low as 500 feet then later "chain home very low" detection to 50 feet) an early warning system. An early development (1935 by Sir Robert Watson-Watt), it was a fairly primitive but effective radar system that gave a read-out of incoming enemy aircraft's height and range, which was then interpreted by the radar operator and the read-outs telephoned to the sector controllers of our aircraft. I found out that these aerials, three in-line clusters, with masts 360 feet high, were spread at intervals mainly around our South and East coasts during the war. The stations were operated by the RAF, and Ventnor's installation was called RAF Ventnor Home Chain Radar Station.

Ventnor was a most delightful place, and I remember that we stayed B&B at the Sea View Hotel. One enduring memory was how light it stayed till late in the evening – 10.45pm (to give the farmers more hours of daylight the clocks in summer were advanced two hours ahead of Greenwich Meantime). I persuaded my father to climb up to the top of this hill as I called it. The correct name was St. Boniface Down, and it was about half a mile ENE from the town and 791 feet high (241 metres) – such a steep climb that my mother declined to join us! We climbed to the top and found a small deserted hut next to the aerials. On walking around, I saw bits of twisted metal that looked like aircraft parts. One piece had on it the Nazi Swastika, definitely part of a wing, so there it was – a crashed enemy plane. Curiosity had got the better of me; on further investigation, poking out of the ground I found what turned out to be a very large shattered engine, obviously from the plane. The radar station had been raided by Stukas in 1940, but the attack was only partially successful after they were counter-attacked by Hurricane aircraft. The RADAR aerials were left intact.

I had a very good holiday, walking with my parents and playing in the sea front paddling pool, in the centre of which was a model of the island. I paddled there, and played with my model yacht, and also paddled in the sea, played on the sand and looked for crabs in rock pools: in the light summer evenings walking along the cliffs. Many, many years later, having obtained my Private Pilot's Licence, I flew my Cessna G1000 182T Skylane back to the island, landed at Sandown and then went by road to Ventnor. The happy memories came flooding back when, much to my surprise, I found that very little had changed on the sea front and that the paddling pool I had spent so many happy hours in was still there, looking just the same, if a bit more jaded. I visited what was the railway station to find it a builder's yard and industrial buildings. The stationmaster's house was still standing. The railway tunnel from Ryde pier to the station passed through St. Boniface Down and was 1,312 yards long (1,200m) the station itself lay 294 feet (90m) above sea level. The tunnel now used by Southern water. An easy walk to the town and a tough climb back up to the station.

When I was ten years of age, my parents gave me a red Raleigh bicycle – my pride and joy – subsequently I installed dynamo lighting for front and back which I rode everywhere on the main roads. On some days, later when I was at Ealing Grammar School, I rode to school along the then-narrow A40, taking about three quarters of an hour – an unthinkable journey now with the incredible volume of traffic and three lanes each way! After school, I would ride to Greenford to visit my maternal grandmother and my cousin Maureen, or take the push-pull train from Ealing Broadway to Greenford Halt. Thursday afternoons were sports in all weathers, summers and winters, on a field owned by the school at the end of my cousin's road – most convenient. Their house backed onto the Greenford-to-Ealing branch line and we used to play on the embankment looking for grass snakes – they were there a-plenty. My mother's brother Uncle Eddie was stationed in India during the war, and he had a wonderful and colourful butterfly collection, pinned neatly in glass-fronted cabinets, collected during his time there, which fascinated me. He had a medium sized greenhouse, heated summer and winter by water pipes warmed by a small anthracite stove on the outside of the house. He grew a number of exotic flowers and plants in there – I guess a hangover from his days serving in India.

While waiting to take the push-pull branch line train back to Greenford underground station – then a Central London underground train (this was above ground at that stage of the line) to Northolt – some of us boys travelling together would place pennies on the line at the halt to be run over by train wheels and squashed flat. This may seem trite now but then it was good fun and had a frisson of danger! On winter evenings when darkness fell early the train carriages were lit by gas lamps! At other times I would cycle to Denham village down the old A40 with friends (as I mentioned above, it was quite safe then with relatively little traffic), where there was a small stream, and we would catch minnows, getting thoroughly wet and dirty in the process, place them in a jam jar, and take them home. Near the village was a very large house, standing in what appeared to be huge grounds, surrounded by a red-brick perimeter wall – it looked very foreboding.

Many years later, I discovered the property was the home of famous film-maker and producer Alexander Korda (died January 1956), who built and owned Denham Film Studios, not a stone's throw from his home. Little did I know at the time, that a few decades later I would be heavily involved in the same industry, owning recording studios with my partner Johnny Pearson and recording scores for many major motion pictures and television series! I used to visit my other cousin (my paternal uncle Tommy's daughter Sheila) in Hove, where Uncle Tommy had his own business manufacturing lampshades. We were about the same age, and her mother – my Aunt Nancy – worked in the food industry, so no food rationing for the family in Hove! We used to exchange one-pound bags of sugar for sweets at the sweet shop! Didn't do my teeth much good later on – copious fillings! Brighton and Hove, in those days, was a wonderful place to visit, although some of the beaches had the war defences on them and the West and East piers were not usable, as the middle sections had been cut out to deter any German invaders, as was the case with all piers around the coast. We thoroughly enjoyed ourselves nevertheless! My parents told me that if I passed the Eleven Plus exam they would buy me a watch, and after I passed they kept their promise. I was given a Bravingtons Renown watch, engraved on the back "from Mum and Dad June 1949". I still have it, but sadly it is no longer working. I was promoted, if that is the correct word, to Head Choirboy, the result of which was that I had to sing all the soprano solos! On Sundays, there were two services a day – 11.00 am Matins and choral Evensong at 6.00pm. The best part was singing at weddings on Saturdays, sometime as many as three or four in a day, or at the occasional funeral. As Head Choirboy, the church Verger paid me the money that he'd obtained from the Best Man for our singing labours. I would be given the correct amount of money to pay the other boys, often a pound or a ten-shilling note, or even more depending on the number of boys singing. The money had to be split between all the boys

(£1 in 1950 equals £31 today), and the summer months were particularly good for income! We all bicycled down to the post office where I changed the money and shared it with the others, each getting an equal share of two shillings (10p = £3.16 today) or half crown each (12.5p = £3.95 today). Not only was this useful extra pocket money – it was also my first experience of earning a living from music! We often saw the mother of the bride crying during the ceremony and as young boys found it amusing. We saw all sorts of different people getting married, and on one occasion the best man dressed shockingly casually for the time and even chewing gum! Boys can be cruel, and we thought and said that all these people only did Christenings, Marriages and Deaths – just three visits to church during their lifetime and their last one hardly counted because they'd know nothing about it, despite the fact that he or she would be there in body, if not in spirit! Although funerals are generally solemn affairs, I did sing at some that were sad for the family involved and at others that were a true celebration of the deceased's life, with plenty of humour. We were paid for our efforts when required to sing, although we weren't often needed. As head choirboy, I was expected to do my share of solo anthem singing. I frequently sang solo works at Matins or Evensong services and (hopefully) I sang without too much vibrato. I did not have perfect pitch but relative pitch. The church had a fete in June each year in the Harrow Hospital nurses' home grounds, which were pretty extensive – although it has long since become housing. Myself and a friend would join train sets, combining our Hornby 00-gauge train sets in the vicar's garage and then charging an admission fee, the proceeds of which were contributed to good causes. A bunch of us young boys had girlfriends – and mine was Pat Steele. Her father did not approve – he was very strait-laced and strict with Pat. Regardless, Pat and I got on very well and spent many happy hours together. We eventually lost touch. My father liked going to concerts, and when I was aged ten he took me to a solo piano concert in a school hall in Ealing. The piano, I noticed, was a very large Steinway Grand. On the left hand side of the hall were a number of rooms that you could see into through large glass windows, and I discovered that these were the very well equipped school laboratories. The school was closed for the holidays, and I plainly recall how keen I was to attend that school if I passed the Eleven Plus – it had a good feel. I have never forgotten the smell of the newly treated, dark-oiled pine, herringbone-patterned hall floor. My class teacher was confident we would pass the Eleven Plus but nevertheless said if we did she would eat her hat. After the results, when most of the class passed, including me, she came in with a chocolate hat that she had made and promptly ate it – she kept her promise! My chosen school was Ealing Grammar School, at which I was given a place. Back when I attended that piano concert I'd had no idea I would eventually attend there. At age 11 in 1949, I was now dressed in long trousers – hooray! I duly reported to the school, which had a good reputation and a disciplinarian headmaster, Mr Sainsbury-Hicks – a formidable man to us boys, but I suppose he was all right to the masters. On the first day I felt out of my comfort zone. I recall Sainsbury-Hicks telling us he would be visiting America for three weeks. The reason I cannot recall, except that when he returned he spoke with an American accent, which greatly amused us. Only three weeks to develop an accent! He soon reverted to English though! Our masters were excellent teachers. One abiding memory I have is of our French master, who was another formidable man, although his name is lost in the mists of time. We had to learn the French verb tables by heart – and there were a lot of them! He was the only master whom I did not like and the homework was an utter bore, mostly spent writing out the French verb tables. We boys were very wary of him. Some of the boys misbehaved, as boys do, and had the cane – six of the best! Fortunately, I kept my head down and out of trouble… well mostly. Classes were divided into 45-minute slots with a different master (no female teachers at the boys' school!) for each subject. They all wore gowns and, on special days, wore the full-hooded gowns that indicated from which university they had graduated. Other lessons of a practical nature were woodworking, which I enjoyed, and learning to swim at Ealing swimming pool, PT (physical training) and best of all skiving off on some afternoons with my mate Geoffrey to visit the cinema – we never got caught despite having to pass the headmasters office window! How lucky can you get! I regret to say that I should have worked harder, and my homework was sometimes done hurriedly in Walpole Park Library at lunchtime to free me up in the evenings. My early interest in music was coupled with my love for electronics and audio amplifiers, which I began to build at age twelve, and so my evenings after tea were often spent building valve amplifiers in my bedroom. I also built a crystal set, listening to a weak radio signal on headphones, and had an old valve radio in my bedroom on which I listened to Radio Luxembourg on 208 metres and searched for other stations on long and short wave. I bought components and circuit diagrams for the amplifiers from a radio shop in Harrow with my pocket money and choir earnings. Some amplifiers I could not get working to my satisfaction but when I did I listened to records in my bedroom loudly! West Coast American Jazz on the then-new 10 inch (mono) 33⅓ LPs, particularly a group of West Coast studio musicians, featuring Barney Kessel (guitar), Herb Ellis (electric guitar), Marty Paich (composer and pianist) and Shelly Mann (drums – years later I would work with him at Lansdowne). They were all great players, hence my love of jazz, which continued later when I met producer Denis Preston at IBC (The International Broadcasting Company – see Chapter 2).

School authority was good: no talking in class or back-chat, although we did manage to cause a few problems! The desks were ancient – no doubt pre-war or even older, they were of the single-sitting type with a lift-up desk top covered in penknife name carvings or ink doodles by boys who had sat there previously. They had an inkwell set into the top of the desk on the right hand side. Opposite the school was a chemist's where we bought calcium carbide, (all the chemicals one could buy were kept in small wooden draws and carefully labelled) which was properly used mostly in cycle lamps for night-time illumination. A prank was to drop chunks of it into the full inkwells, so that the ink bubbled from the released stench of acetylene gas and overflowed as if it were going to explode. In our young naivety we did not realise how flammable the released acetylene gas was! The stench was awful and no one owned up: "Not me sir!", "Not me either sir!" and so on… Somehow, no punishment followed! During the lunch breaks when no homework beckoned, I would walk in Walpole Park with a school friend, Geoffrey Newbury, after buying a quarter pound of powdered sulphur from the same chemist, along with a quarter pound of sodium chlorate – we had a good chemical lab at school. We would mix a very small quantity, and I mean SMALL quantity, of the two chemicals, place the mix on the ground with a stone on top and stamp on it. The resultant bang was very loud and people looked round – but no parky (park keeper) came to investigate, so we got away with it!

During the school summer holidays of 1950 and 1951, I had the opportunity to accompany my father to British Telecommunications Research (BTR a subsidiary of Plessey Electronics) Taplow Court, Taplow, Buckinghamshire. A large country house set in about 200 acres with grounds stretching down to a sleepy backwater of the Thames (close to Cliveden the former home of the Astor Family and now synonymous with the Christine Keeler of Profumo scandal fame in the 60s). In the laboratory where my father worked they were building a small experimental digital telephone exchange. My job was to test the Hivac miniature wire-ended triode valves to match to a close tolerance – and the ones that fell outside specific measured parameters were not used. It was a simple test, involving measuring the cathode, plate and grid currents for set voltages, and any one valve might have differed from the manufacturer's published average characteristics. These miniature Hivac triode valves were used as an "on or off" switch, depending on the bias condition, to correspond with noughts and ones – my first introduction to the concept of digital. The 700 rpm rotating magnetic-coated drum was about four inches in depth and had a diameter of twelve inches, with a series of read/ write heads vertically stacked around it – the small heads were wound in the lab. One of my father's tasks was to dial every single London telephone number – a massive task because there was no automatic testing program then. I heard no more about the project – it must have bitten the dust or been overtaken by advances in computing technology. That's research for you!

One advantage for me was that the metal workshop made, to my requirements, aluminium welded chassis for my amplifiers at no cost, into which I cut the valve-holder holes and any others required for potentiometer and transformer fixings and so on. In the same lab where my father and I (in the holidays) worked, was a side room with a locked door. Being nosey, I asked what happened in there – the answer: "Secret government work!" What mystified me was every now and again a chap would come out with a potentiometer in his hand and make a resistance measurement of it. "How odd! Don't they have meters in there?" I thought. I guess he was verifying the calculated measurement result made in that room, but I asked no further questions.

The famous Ealing Film Studios, with Michael Balcon at the helm in those days, were next to the school and on one occasion they needed extras, paying three days at fifteen shillings a day – good money then, and today worth around £21.00. The film was *"Lease of Life"*, starring Robert Donat. This was shot in Eastman Colour, and was the only film he did with the Ealing Studios (now owned by the BBC) and proved to be his penultimate picture – he was not in good health. The lighting set-up up took so long and the lights were powerful carbon-arc models with smoke from the arc rising up to the high roof ceiling. The ventilation in the roof appeared to be good, but I wondered about this smoke. I later discovered these noxious fumes consisted of carbon dioxide, carbon monoxide and oxides of nitrogen. Carbon-arc cinema projectors did at least vent their fumes via an extractor to the outside world. Our scene was a sermon in a cathedral, which was a set built on one of the huge sound stages; it was my first experience of playback, during which we had to sing the hymn to the backing track. I was captivated by how long it took to set up each shot, particularly the lighting, to ensure there were no unwanted shadows on set.

During the middle 50s I decided to learn ballroom dancing in order to meet a few girls! There was a dance studio over Burtons the tailors in Harrow where most of the 78s played were Victor Sylvester's strict tempo dance records – he was on the Columbia Label. To learn the quickstep, the favourite record was *"Mr Sandman"* (not by Sylvester, by the Chordettes a 10" shellac mono 78 rpm released on the Columbia EMI label) played over and over again while the dance teacher repeated over and over again, "slow, slow, quick-quick-slow" – that rhythm applied only to the Quickstep and Foxtrot. Sylvester's records were indeed in strict beats per minute. I left after a few weekly lessons. And although I did manage to meet a few girls, the hormones ruled and we changed partners like ships passing in the night!

And so it came to pass after taking the O-level examinations that the result of my not working hard enough (too busy thinking about amplifiers and music!) meant I only achieved one O-level pass in woodwork, and failed miserably in the other papers. I cannot even remember the subjects now. I therefore did not make it to the Sixth Form, and the school wanted to put me and some of the other boys who had also failed miserably into a form called Form Five Remove – it was a humiliating move and, for me, a distasteful one. On reflection the school had our best interests at heart, the idea being that we could have another go at our O-levels.

I decided to leave school at fifteen years old, in 1953, with the blessing of my parents. Fortunately, while still at school, I had obtained a Saturday job at the South Harrow music shop through my friend Trevor Clark, whose father owned it – and a very successful one it was too. Trevor Clark had a sideline showing cartoon films at children's parties so on some Saturdays when for some unexplained reason Trevor couldn't do it, I was asked to go along and show 16mm Walt Disney cartoons on a very old 16mm portable projector with dreadful optical sound! I had to cart the projector and its large speaker and the long heavy speaker cable around by bus – a hell of a job because it was all so bulky and heavy. I got paid 5 shillings a job (two half crowns = £7.02p today). In the end, I refused to do it anymore, as I was expected to go everywhere in the area by bus carrying all that heavy kit. That said, the extra money did enable me to buy more records at cost price from the shop!

Most weeks during the school holidays I went to a Harrow junk shop where they had many second-hand records, I chose the best ones (mostly 10-inch 78s in good condition – not over-played, no grey grooves, as these were rejected) by what I thought were good artistes on various labels – HMV, Parlophone etc. I paid sixpence (70p) for each one, then cycled to South Harrow market to a market trader, and sold them on for one shilling (1.40)! Not a bad return. This I did most weeks, sometimes twice a week, depending on what the junk shop had in stock at any one time. I was already exploring career opportunities in the music business!

After I left Ealing Grammar School in 1953, my father decided that working for British Telecommunications Research in the laboratory at Taplow Court, was a laborious dead-end job, and decided to change direction and buy a post office-cum stationery and sweets shop at 54, Fulham Palace Road, in the Hammersmith area of London. Much to my dismay, this necessitated selling our home in Northolt. The move did not appeal to me, with the trolley buses and constant traffic rumbling past the shop. We continued to live over the shop for a number of years, and were still there during my time in National Service, and although I never liked living there very much, I put up with it on my occasional weekend leave. Eventually, much to my surprise, my parents having made a success of the shop by working long hours, then decided to move to a bungalow in Northwood Hills and buy a similar shop in Hayes, Middlesex that had a three-man hairdressing saloon at the back. The hours were long, but the shop had the advantage of being opposite EMI Hayes so there was plenty of business. This move could not have been better timed, as it took place in the last few weeks of my National Service and so I could return home to the new address.

When I left school, I realised I wanted a taste of the real world and therefore had an opportunity to join full time the South Harrow music shop, next to South Harrow Market, which had now been sold by the Clarks to Ken Crossley, a guitar player in Sidney Lipton's band. It was just past my fifteenth birthday and it was no longer a part-time Saturday job – in fact I was eventually made responsible for managing the store, and ordering stock from EMI Hayes and from another supplier in Tottenham Court Road, London, as Ken was often on tour with the band. The ten and twelve inch "shellac" (shellac is a natural resin – a brittle material) 78s records came in cardboard boxes of 25 from EMI. On many occasions they had to be returned because they were "dished", where the record press operator had sleeved them and placed them in the box warm (the pressed record was partially cooled in the press) and stored flat, thus the weight of them caused them to "dish". Unplayable. The box should have been stored on its side to prevent this happening or the records left to cool! Thanks EMI Hayes...

The shop was usually very busy, especially the day after Boxing Day when we were rushed off our feet with people spending their Christmas money! In Ruislip there was a USAF base and on their payday many of the Americans came to the shop and bought mostly Hank Williams ten inch 78s by the boxful – those USAF guys certainly had money! By contrast, every Thursday there was a gentleman who came by the shop – always by bicycle, and dressed in his working clothes with remnants of wood chip on them and emanating an overpowering smell of fresh wood. Being curious, I asked him what he did in wood. He told me he was a coffin maker in Harrow! I didn't go into any further detail with him! He spent good money and the thought went through my mind there must be good money in the coffin making business – after all you would never be short of work! After each visit he got back on his bike and rode off to have his lunch at home, a regular weekly routine. Sometime after Ken bought the record shop, he installed three booths for auditioning 78 and 33⅓ rpm records, with small low output valve amplifiers driving small speakers. The resultant sound was awful but it didn't deter customers – this was a time before Hi-Fi took off! Some punters came in to listen on the pretext of buying, listened to a record or more and then walked out! We sold mostly 78s (this was before the vinyl 45 RPM or 33⅓ RPM LP became the norm) and the shop was exceptionally well stocked with 10" and 12" 78s and some of the early LPs. We stocked popular American big band 78s by Count Basie, Duke Ellington, Woody Herman, Stan Kenton, Benny Goodman and so on as well as the likes of the Ted Heath band, while classical records usually came in the form of 12" 78s, mostly by order only. I remember very well, the big hit shellac 78-sellers such as Kitty Kallen's *"Little Things Mean a Lot"* and Frank Chacksfield, *"Little Red Monkey"*, which featured a very strange electronic sound that I later found out was a clavioline. And then there was Charlie Chaplin's theme *"Limelight"* and *"Ebb Tide"* with sea gulls on the intro and the end. Little did I know then, that ten years later I would be working with the Chacksfield Orchestra! Other records I recall included Mantovani's *"The Song from the Moulin Rouge"*, and Mantovani with David Whitfield's *"Cara Mia"* – a string sound the like of which I had never before heard: what recording technique was that? Much later I discovered the arranger was Ronnie Binge, and Mantovani became well known for that string effect, purely in the scoring, which was referred to as cascading strings or the "Mantovani Sound". He was on the Decca label. There were the usual releases of Christmas records from all of the record companies, but the seller I remember best was Bing Crosby's *"White Christmas"*, which was hugely popular – we sold hundreds! These were the early days, before "Rock 'n Roll" became established, and an era of change was fast-approaching the record industry and musical genre.

Chapter 2
First Studio Experience – IBC Studios

In late summer 1953, Ken Crossley announced he was recording with Sidney Lipton's band on a regular gig at IBC studios. The name meant nothing to me, but he invited me to the recording; I later learnt that it took place in IBC Studio A. This was my ever first time in a recording studio, and I was fascinated by all the equipment. I was invited to sit in the control room next to the engineer, and there was one large speaker that was, I thought, quite loud but had a very impressive sound. I had heard nothing like it before – I later learnt this was a Lockwood monitor loudspeaker, fitted with a 15″ Tannoy dual concentric speaker.

Monitoring was standard throughout the studios, with the exception of the assembly room, which contained one Lockwood monitor speaker and a corner horn-loaded Voigt speaker, and very good it was too – an excellent sound. The Voigt had a wide frequency response. Stereo was not around in those early days at IBC. We boys used to connect a sine wave signal generator through the equipment and crank up the frequency to get to 25 KHz, listening for when the signal dropped off – good young ears! It went to show how good the broadband valve equipment was in those days, a tribute to the guys in the workshop and their design skills.

The assembly room (today we'd call it an edit suite) was equipped with three BTR-2 tape recorders (a three-head machine: erase, record, playback) and a small console and was mainly used to edit the studio recordings and compile the Radio Luxemburg shows that were recorded by IBC. Back in the session, sitting next to the engineer at the control desk, I could watch closely while he constantly twiddled the rotary knobs. As a first experience in that environment, I thought the band played well and the engineer recorded it well – I had the live sound as a reference because I initially listened to the band on the studio floor. During the recording, I had a pleasant chat with the engineer between the takes. He asked me if I was interested in this type of work. I told him yes I was and explained something of my background. To my astonishment he said they were looking for staff and he would contact me. That person was Allen Stagg whom I later learnt was the Manager of IBC studios. He joined the company in 1952 from Radio Luxembourg.

I waited in anticipation for several months for the call to come but it never did so I decided on a bold move to call him and remind him of what he had said. He kept his word, and immediately asked me to come for an interview. I jumped at the chance! I duly arrived at 35 Portland Place, located near the BBC, at the appointed time looking my best (at my mother's insistence) and was shown to his office via an old lift to the third floor. Allen Stagg questioned me about my background and why I wanted to come into the recording world. I cannot exactly recall my answers but must have said something he liked. I was offered a job, there and then, five guineas a week (£5.25 – £130.92 today): the hours to be Monday to Friday 9.00am to 6.00pm and on Saturday 9.00am to 1.00pm. Little did I know then the kind of hours worked with sessions! I have much to thank Allen for, and I would never be where I am today if it hadn't been for him. However, when Stagg was interviewed by John Repsch in April 1983, he apparently thought that I was "lacking in talent"!! Sadly, Allen passed away on August 29th 2014 while we were exchanging emails about IBC history.

I joined IBC in May 1954, as my first studio job, nine years after the end of the Second World War. After nearly six years of war the ending of hostilities in May 1945 left the country almost on its knees. In fact, it was not until July 1954 that food rationing finally came to an end – some 14 years after it was first introduced! Difficult, perhaps, for the reader to comprehend now. The "feel good factor" of the Festival of Britain in 1951 was supposed to be the start of a new beginning. As a schoolboy, I went to the Festival of Britain on the South Bank of the Thames, an interesting event with many outside exhibits, including the Dome of Discovery and the peculiarly shaped cigar-like Skylon pointing skywards – it appeared to float, and was an odd-looking piece of design. The Festival Hall was built for that event and there was much publicity about the hall acoustics being the same empty as full – the hall seats, when up, were absorbers designed to ensure the acoustics did not change, irrespective of how many people were in the audience. The 7,866-pipe organ was installed later in 1954. It took four years (1950 to 1954) to build and the largest Pedal Organ Pipe (Principal) was 32 feet long, giving a fundamental tone of 16.4Hz – the pipes in the 32ft octave sound notes that are lower than any pitches on a piano or any that an orchestral instrument can produce. The space in the hall could cope easily with that long wavelength. The organ was designed by Ralph Downes and constructed by Harrison & Harrison Durham and later, when I visited the Festival Hall, it sounded fantastic. Having had a long interest in organs since my chorister days, I was particularly impressed by the frequency range of the instrument, most especially the thunderous pedal organ.

IBC Historical Background – Radio Normandie (Normandy)

IBC Studios held such a pivotal position in the early days of the recording and broadcast industries that it is worth exploring its background in some detail. As a studio, it was the spawning ground for much innovation and talent, and a forerunner in terms of the business models that it helped pioneer.

Captain Leonard F. Plugge, an ex First World War RAF pilot, was an entrepreneur with a passion for radio and cars and also Conservative party member of Parliament from 1935 to 1945 for Chatham, a division of Rochester. He was an enthusiastic radio man and installed numerous radio receivers over the years while touring; a Western Electric superhet and latterly a Standard Telephone nine-valve radio (valves in those days) in his car, thought to be at the time one of the first car radios, with a large rotating frame aerial mounted on the offside. He loved touring Europe in the car listening to many European stations. His father was Belgian, so the family name was pronounced "Plooje" which his father insisted was properly pronounced when in France and Belgian, and not "Plug" as the English pronounced it!

The family lived in London at 5, Hamilton Place and he founded The International Broadcasting Company (IBC) on 12th March 1930, with the objective of providing an alternative commercial broadcasting station for British audiences – a rival to the staid British Broadcasting Corporation. The offices were in 11 Hallam Street, near where the then-new BBC Broadcasting House was being constructed. Now the redeveloped BBC calls it New Broadcasting House for the digital age.

IBC soon built up a radio station network buying airtime and making money from the advertisers. Its programmes were broadcast from various radio stations in Europe in addition to Radio Normandy, Radio Paris (which closed down and subsequently became Radio Luxembourg), Poste Parisian, Radio Toulouse, Radio Côte d'Azur – Juan-les-Pins, Radio Lyon, Radio – Eire, Radio Athlone – Eire, Radio Madrid, Radio Barcelona – Spain until the Spanish Civil War ended the broadcasts.

IBC transmitted the pre-recorded programmes at various hours of the day. They also worked indirectly with Radio Luxembourg, and Plugge sold airtime as sponsored English language programming planned for audiences in Britain and Ireland. He sold the schedules of the various radio stations he visited to the BBC to be published in the *Radio Times* and other magazines!

Peter Harris, technical engineer at IBC: *"Plugge had no ownership connections with Radio Normandy. In 1933, he had hoped that IBC would get the sole concession for the new Luxembourg station's English broadcasts. However, it went to Radio Publicity (London). His only concession was being able to print Radio Luxembourg's listings in his publications."*

Radio Fécamp – the Birth of Commercial Broadcasting

Radio Fécamp was a small commercial station established in November 1926 based at Villa Vincelli la Grandier in Fécamp on the Normandy coast.

The owner was a M. Fernand Le Grand whose principle job was a director distiller of Benedictine, but he was also a passionate wireless amateur. Le Grand had a small radio transmitter in his drawing room, which he used to sell shoes by way of a radio programme over the air, with advertising for a shoe manufacturer in Le Havre. It was not a powerful transmitter, with 50 watt station broadcasting on 246 metres (1220 kHz), but it reached an audience and the shoe sales increased!

Plugge, owner of The International Broadcasting Company (IBC), did a deal with Le Grand after discovering the small station while touring France. Plugge offered to buy airtime to broadcast programmes in English for audiences in Southern England. On 6th September 1931 the first broadcast was made from Le Grand's house. Eventually, studios were set up in the hayloft of the old stables and the transmitter power was increased to 10kW. This higher power meant reaching larger audiences and with advertising as the revenue source it was very successful and made Plugge a lot of money.

In November 1935 a new more powerful 25Kw - broadcasting on 269.5 metres (1113 kHz) - transmitter was built at Louvetot, 20 kilometres South of Fécamp along with new studios at Caudebec-en-Caux. The English programmes were delivered to the station by Plugge, who was by now part of the management. Radio Fécamp later changed its name to Radio Normandy (broadcasting on 274 metres.) and, thanks to the new transmitter, its broadcasts could be heard all over Southern England.

Leonard Plugge, Lenny as he was known at IBC, had successfully demonstrated that state monopolies such as the BBC could have a serious competitor. Radio Normandy played commercial records with advertising in between unlike the BBC, which only offered the British people programmes to inform, educate and entertain.

One of the earliest sponsors on Radio Normandy was Spink and Son Ltd (purchasers of unwanted jewellery) and Filmophone, manufacturers of a roll-up record that could be delivered in a cardboard tube. HMV and Broadcast Record soon followed suit with similar products (Source: *And the world Listened* by Keith Wallace pg. 82 Kelly Publications 2008)

In March 1938 the wavelength was changed to 212 metres then in December 274 metres 1938, the programs were mainly recorded on discs at IBC's London studios other discs from America, and flown out to Luxembourg, featuring 15-minute sponsored shows. Live programs also went out from France.

Before war was declared against Germany on 3rd September 1939, the orchestral segment of *Keep The Home Fires Burning* was used as the close-down theme for IBC Radio Normandy. IBC tried to revive the station during the British expeditionary forces time in France, but with all American programs and news bulletins read by Bob Danvers-Walker – later the voice of Pathé Cinema newsreels and the presenter of some of the shows I was recording with Joe Meek. One such show was *People Are Funny*, recorded from venues all over the country. *Shilling A Second,* a radio quiz show with compere Gerry Wilmott and announcer Patrick Allen, was frequently recorded in Conway Hall in Red Lion Square, London, where IBC had a small permanent control room recording on a BTR 2 Transportable (mono ¼") Recorder.

IBC worked indirectly with Radio Luxembourg until the outbreak of World War Two, when the Germans took over the station, but the war meant that the stations that Leonard Plugge worked with would go dark between 1939 and 1945, overrun by the German armies. On the day of the Nazi invasion, Radio Luxembourg installations were among the first objectives of the Wehrmacht in the Grand Duchy. Four weeks later, German troops restarted the station, using it for their own communication until October 1940, when the transmitter was incorporated into the Reichs-Rundfunk-Gesellschaft and used for Nazi propaganda. Radio Luxembourg was used as a relay station by German stations such as Radio Hamburg to broadcast Nazi propaganda to Britain from 1940-1945 by Irishman William Joyce, better known as Lord Haw-Haw. (Source: Radio Luxembourg history: http://www.radioluxembourg.co.uk/?page_id=2)

The BBC

John Charles Waltham Reith (later, 1st Baron Lord Reith) was appointed General Manager of the BBC on 14th December 1922, and became Director General in 1927 when it became a public corporation. Under Reith, the BBC provided quality conservative programming but listeners began to find this type of programming dull. This was especially true on Sundays, when broadcasting began at 12.30 pm to allow listeners time to attend church, and the BBC's output was of a religious nature, complemented by classical music and serious programming – for Reith it was a day of rest. Reith gave the public what he thought they wanted, and not necessarily what the public actually wanted. To listen to Plugge's commercial radio was a breath of fresh air and it was also making money – huge amounts of it! Plugge was deemed to be a thorn in the side of the BBC; it appeared that Reith's cage was surely rattled. Commercial radio stations had a receptive audience.

As a matter of interest, Radio Luxembourg was a bigger concern to the BBC than the other commercial radio stations in Europe, as it owned a very powerful transmitter. At 100 kilowatts, only Radio Moscow's was more powerful at 300 kilowatts. The British government was particularly hostile to the station because the signal received on UK soil was loud and clear. There were a number of frequency changes before the station was eventually permitted by a European Broadcast Conference to make a final change to 208 metres medium wave (1442 kHz). Their English language programmes were originally broadcast on 1250 metres (2398 kHz). These new commercial programmes transmitted from continental soil were a threat to the BBC. Such was the threat to Reith's organisation that he appealed to the French government to ban Radio Transmissions from Europe, claiming the stations were in contravention of the BBC broadcasting charter! There was talk of a ban by trying to negotiate an agreement, it came to nothing.

Radio Normandy already had gained a bigger audience in southern England on Sundays than the BBC! IBC Radio Normandy was closed down six weeks after the invasion of France when the Germans took over the station. Hitler managed to do what Reith could not!

Now there was no opportunity for commercial broadcasting to be available from inside the UK the BBC held a monopoly. Under John Reith it was deemed by many to be a very stuffy station when founded in 1922 – called 2LO, the London station of the British Broadcasting Company – and it stayed that way until the '60s, when the face of broadcasting changed forever with the establishment of BBC Radio 1, Radio 2, Radio 3 and Radio 4 in September 1967. These were the BBC's response to the threat from the growing popularity of the pirate radio stations.Radio Caroline was the first – and very popular with the younger generation, much to the annoyance of the onshore broadcasting authorities, the BBC included. They were located outside UK territorial waters – usually transmitting from old rust-bucket ships.Radio Caroline came on air on Easter Sunday 28th March 1964. Other offshore radio stations followed such as Radio London, Radio Atlanta, Radio North Sea International to name but a few. Now we had commercial pop music broadcasting and whether the records played or not was no longer at the sole discretion of the BBC who had their own record company-preferred playlist. The record companies' control of popular music broadcasting in the UK and the radio broadcasting monopoly by BBC were outwitted. Unlicensed by any government, Radio Caroline and others became formally illegal in 1967 through the Marine Broadcasting Act. People wanted rock 'n' roll; they wanted a change. The BBC was conservative and slow to react to the changing tastes of the nations – especially those of the youngsters.IBC's original offices were at No 11 Hallam Street, near BBC Broadcasting House in London, and were the London end of Radio Normandy. They moved to 37 Portland Place (although some records say No. 36), and this building, adjacent to No. 35, was taken over by a British weapons secret development unit during the war. IBC subsequently moved in to No. 35, which is where I worked. No. 35 was originally a house built for family occupation, designed by the architect John Nash. Many parts of the building retained the original doors, and the Studio "A" plasterwork ceiling had the original ornate decoration, as did some other parts of the building's ceilings. When I joined, the original large control room on the ground floor had an outstanding white marble Nash fireplace. The control room later moved to the mezzanine floor where its control window looked down on the studio floor.

The London Recording Scene in 1954

When I began work at IBC in 1954, there were only a few other studio facilities in London. IBC was considered to be the largest independent. Among them was "The Columbia Gramophone Company" (HMV), which was originally established from the 1931 merger with the Graphophone Company. The new company was called Electric and Musical Industries, or EMI as it became known, with the company's recording studios situated in Abbey Road, where for some years the words HMV were written over the top of the entrance doors. HMV Studios opened in 1931. Today it is better known as (EMI) Abbey Road Studios. London's other recording studios at the time included Decca (West Hampstead), Levy Sound Studios (New Bond Street), Star Sound, the BBC Studios in Portland Place, and BBC Aeolian Hall in Bond Street, plus Radio Luxembourg Studios at 38, Hertford Street. Leonard Plugge eventually sold IBC in 1963 to brothers named Berman and, according to Allen Stagg, they were unpleasant people to deal with, *"...knowing nothing of the business – not interested in it, they were asset strippers".* The company was sold again to orchestra leader Eric Robinson and musician George Clouston, then bought by Chas Chandler in 1978 and renamed Portland Studios.

Leonard Plugge eventually died in California aged 91 years.

Working at IBC

Having never previously been in a working studio environment like it I was initially overawed by the size and technical capabilities. I discovered the workshop in the basement, with white tiles covering all the walls, amid a maze of rooms – very Victorian. This is where people worked, built the studios' recording consoles maintained the tape machines and repaired other pieces of studios equipment. In those early days, there were no commercial manufacturers of professional equipment. There were, of course, very few studios around to warrant a recording console manufacturer setting up in business, and the BBC, for example, designed and constructed its own gear, as did IBC. The design and reliability of the equipment, using valve (tube)-based electronics, depended largely on individual studio requirements and, especially, on the ingenuity of the technicians. I subsequently learnt their names: Peter Harris, Ian Levine and, later, Dennis King. In those early days, equipment for the studio, apart from the BTR tape machines, was usually designed and constructed in-house by these talented engineers. (In this picture the two techs, Peter Harris and Ian Levine, are building a new console for Studio A). All of us young newbie's had to wear white coats and were general dogsbodies, although we were supposed to be trained! We mostly swept the studio floors and collected the dog ends, but more interestingly also got to set up the studio layout and place the microphones. We had to follow instructions from the engineer allocated to the session with the details usually hand-drawn on a scrap of paper or not at all – often just verbal, and in that case we usually knew what the engineer required because many were continuation sessions. Sometimes we also cleaned the loos, especially on Saturday mornings, when there was not much else to do. Many musicians were very heavy smokers and after a three-hour session the studio was thick with cigarette smoke, no 'elf an' safety then. Brass players were always thirsty, and the pub beckoned in the "tea" breaks. I recall there was no air-conditioning in either Studio "A" or Studio "B" located on the top floor. The studios became very hot with so many bodies in them, especially in the summer. Peter Harris recalls, *"When we had band sessions in the evening, neighbours living behind the studio building often complained about the noise, particularly if the musicians asked for the windows to be opened on a hot night. In the end, we acquired several large fans, which we installed at the front of the studio."*

Another chore that Sidney, a fellow newbie, and I had was to collect the lacquer swarf from Ken Wiles' disc-cutting room – he was always very busy cutting master lacquer 78s for IBC's various record company clients. The common name (although incorrect) was "acetate", this term derived from the material, nitrocellulose acetate, used in cuts before 1934. Today the core material, aluminium, is coated with black nitrocellulose lacquer. The cut lacquer was boxed for collection and transported to the various record manufacturing plants for processing - lacquers were also used as one off reference discs for the artist, band or producer. The swarf from the cutting head sapphire needle was sucked into a container, which we emptied and carted down to the front basement area. Rather than dispose of it correctly, we set fire to this highly flammable material, which went up in clouds of black smoke with one almighty whoosh! Little did we realise then just how toxic the smoke was! The stupidity of youth!

Also in the basement area, lying out in all weathers, I saw two old German Magnetophon tape recorders – all-valve-transistors came later. I asked Peter Harris about these machines, and he commented: *"IBC had two Magnetophons captured from the Germans. I was asked to operate one at an outside location for Sir Thomas Beecham. I had to play back several master tapes of his recordings. The only snag with these recorders was that it was often necessary to spin the take-up reel with your finger on start up – getting it wrong could be a very expensive mistake. Luckily, everything went well and I didn't have to incur the legendary wrath of Sir Thomas".*

Over 50 years later, I was asked by Michael Bauch, of FWO Bauch, if I knew what happened to them. He wanted to get hold of one, but I explained that I had last seen them in a forlorn state at IBC, lying abandoned in the outside basement area about 1955 and was none the wiser as to their fate.

The balance engineers (today, recording engineers concerts for theatres sound designers) at IBC in 1954 were Allen Stagg (manager),

Tig Rowe, Bernard Marsden, Eric Tomlinson, Jack Clegg, Joe Meek (originally employed as an assistant engineer), Ray Prickett and Ken Wiles (Disc cutting (78s)). The technical engineers were Peter Harris and Ian Levine. James Lock (Jimmy) was another white-coated assistant, in addition to Sidney and myself. Studio bookings were handled by Angela Perberdy, with Connie the telephonist in the entrance hall reception. On occasions, we white-coated guys were required in turns to operate the switchboard at lunchtime. Connie subsequently married Peter Harris.

Connie's lunchtime relief was often Mrs Gwen Shaw, Captain Plugge's secretary. Mrs Shaw, a keen weaver, had a flat on the top floor of 35 Portland Place where she housed her loom. It was a popular meeting place of the local Weavers' Circle! One thing that always intrigued me was the office on the left hand side in the entrance hall. I peeked through the door once, when it was ajar, and it looked like it was preserved from the Victorian era, dark with heavy office furniture and huge curtains. Peter Harris recalls, *"It was the sanctuary of Captain Leonard Plugge. It was also home to the Company Secretary, William Wood, who had been with C P for many years and C P's private secretary Mrs Shaw."*

At Christmas, Leonard Plugge held parties for IBC staff. Peter Harris again: *"We still remember the Christmas that the studio staff were invited to his Lowndes Square home for drinks. Champagne was very freely distributed by flunkies. As soon as your glass was empty, it was automatically replenished. It was the first time that we imbibed that beverage and it had a profound impact on us!"*

It did on me too, when I went there as a newbie. Plugge's Lowndes Square home was an impressive place.

We had a studio caretaker and general handyman called Kevin who was quite a character. He lived with his wife and children on the premises in one room adjoining the garage and, when required, drove the company's rickety top-heavy van, located in the garage at the back of the studios. This opened onto Weymouth Mews, and most handily located for *The Dover Castle* pub, opposite IBC's back entrance. Originally built in the mid 18th century, with an attractive wooden frontage, I was told it had held a licence since 1777. The friendly landlady was excellent, as was the food – my only drink then was a bottle of Toby ale. I often enjoyed sausages and mash with the others. It was the preferred watering hole for the engineers – even in wartime and in my time many musicians frequented it! The pub is still there today, although no doubt it has gone through many changes since the '50s. However, if you wanted to know anything, Kevin had the answers – and all the studio gossip.

Sidney was an ex-cinema projectionist, and I remember he had a penchant for Rita – she worked for Carlton Facilities with offices in the IBC building, a company that recorded background music at IBC for use in hotels, lifts and public places. Sidney eventually married Rita. Carlton Facilities' principal people were a Major Desmond Beatt, who had served his time in India in WWII, and Joan Walker, who produced the music using various orchestras in Studio A; I was sometimes on their sessions as the lowly assistant. John Childs, who worked as Carlton's technical engineer, had a small office on the top floor. He was another brilliant technical engineer, who left IBC to join Tig Rowe, Bernard Marsden and John Terry at the newly founded commercial television broadcaster ATV (Associated Television), which started transmissions in September 1955. Peter Harris: "*At ATV, John was credited with the development of the logo that appeared in the top corner of our TV screens, to alert the network that commercials were imminent.*"

I got to know Jimmy Lock well. He was an exceptionally gifted recording engineer who eventually worked for Decca Records recording many famous names including Sir Georg Solti, Herbert von Karajan, Luciano Pavarotti and Joan Sutherland, for whom he was the engineer of choice. He also worked with artistes such as Kiri

Te Kanawa, Angela Gheorghiu, Cecilia Bartoli, Plácido Domingo and José Carreras, and the pianists Vladimir Ashkenazy and András Schiff. Jimmy died in February 2009 and, having known Jimmy for so long the news saddened me – he was a remarkable recording engineer who worked with the greats. His recordings were superb. Another most talented recording engineer who joined IBC after I left for National Service was Keith Grant. Later he became a good friend of mine – what a character. He died on 18th June 2012, on his boat with a beer in his hand – what a way to go! His funeral was a very good celebration of his life held at a pub on the river complete with a jazz band on a river boat well done Keith!

Sidney was eventually sacked by Allen Stagg as his time-keeping was not good. Peter Harris remembers an occasion: *"Sidney was my assistant on the recordings of* Shilling a Second *(for the Co-Op). I well remember an occasion when he dropped off to sleep when he was due to play in an effect!"*

Allen Stagg eventually moved on to manage EMI Abbey Road studios in 1967, and Peter Harris moved on to Cine Tele Sound (CTS) in Bayswater, as their Chief Technical engineer, where the studios were in a transformed former banqueting hall behind Whitley's, 49-53 Kensington Gardens Square. Eric Tomlinson joined CTS in 1959 at Bayswater and was the chief recording engineer. He subsequently joined Anvil Films at their huge Denham Scoring Stage when Anvil moved out of the Beaconsfield studios where he recorded many motion picture scores such as *Star Wars*, a classic John Williams score, in 1977, amongst many others. Anvil Films originally had their Anvil film unit cutting rooms, post-synching dialogue, Foley, projectors and magnetic film machines, and offices at the Beaconsfield studios better known today as The National Film and Television School (NFTS). The summer get-together parties at Beaconsfield in the early 60s were legendary with food and booze galore with Ken Cameron dressed in kilt glad-handing everyone – a great personality!

In summer 1963, when I was at Lansdowne, I went to one such party with Lionel Stevens, Lansdowne's finance man. Stevens imbibed copious amounts and I recall him driving off in an erratic manner back to London! I fortunately did not have far to go – South Ruislip. Being friendly with Eric, I frequently had invitations to watch scoring sessions at Anvil's main scoring stage when they took it over in 1966. This was Korda's old place and the lease had run out. I got to know the guys at Anvil well. Ken Cameron, Ken Scrivener, Richard Warren and Ralph May, whom I knew rather less well. Ken Cameron, being a Scotsman, was quite a character and liked large, and I mean LARGE, tipples of whisky. He kept hefty bar-size bottles in a cupboard in his office, with labels upside down when the bottles were sitting normally but easily readable when the bottle was tipped for pouring a *very* generous glass. I well remember one session I attended when Eric was balance-engineering a large orchestra for a movie. Ken arrived during the afternoon dressed in his kilt, as he usually was, from what was obviously a long liquid lunch. He promptly sat on the sofa in front of the console and fell asleep snoring. Being such a character he got away with it even with clients present! I mention this small story, because later I used Anvil in the middle '60s to dub jingles to 35mm magnetic film for our advertising clients and Ken always made me feel welcome and was generous with pouring out his whisky; there were many occasions when I declined, other times when I didn't! Especially when I was with Eric after he had finished his session! Whiskey Galore comes to mind!

Jack Clegg left IBC, worked for Decca, and then joined CTS in Bayswater in 1963.

Joe Meek was asked by Denis Preston to find another studio to build from scratch outside of IBC, to work as an independent studio with the latest recording equipment. Joe searched and discovered, in Lansdowne House, an old basement classical recording studio later to be called Lansdowne Recording Studios Ltd, which opened after substantial rebuilding works.

I lost track of what happened to Ken Wiles, Angela Perberdey and the others after 55 years. Angela, however, was still at IBC in the early '60s.

A few years later, I joined Joe at Lansdowne in 1959. In the 50s, IBC was unquestionably a fertile ground for sound engineering both technical and otherwise, of all of which I have fond memories. Although Allen Stagg was deemed by some to be a single-minded, some might say difficult, manager but he recognised talent. I respected him. Allen was keen to encourage the budding young engineers such as myself and we were often asked to draw a studio layout for recordings specifying the mics to be used, and then Allen would correct us if wrong – a most useful part of our training. I recall Eric Tomlinson was also always most helpful when I worked with him on sessions. He was an excellent engineer, with many high profile music scores to his name in later years. We became very good friends. He frequently worked recording film scores as freelance at CTS in the 90s. In the early days IBC was one of only three independent recording studios – the others were Levy Sound and Star Sound. We recorded for Phillips Records, Pye Nixa and Polygon – before the companies had their own studios – as well as Denis Preston's Record Supervision jazz label, which was issued on Pye Nixa and subsequently the Columbia label. We also had other independent clients and a number of American clients whose names have been long forgotten in the mists of time. EMI and Decca had their own signed artistes, who were obliged to record in the companies' in-house studios.

We also recorded many artistes for other record labels, including Shirley Bassey, Petula Clark, Ann Shelton, Frankie Vaughan, Gary Miller, Eric Delaney Band, Rosemary Squires, Dorothy Squires,

Marion Ryan, Denis Lotis, Ken Colyer, Edmund Hockeridge, George Shearing, Harry Secombe, Cyril Stapelton Band, Winifred Atwell, Denis Lotis, Lita Roza, Dickie Valentine, Mr Piano Joe Henderson, Lonnie Donnegan, Chris Barber, Monty Sunshine, Johnny Duncan and the Blue Grass Boys, Big Bill Broonzy, Humphrey Lyttlelton Band, Kenny Baker's Dozen, Peter Knight Orchestra and Laurie Johnson orchestra. And many others: an impressive list of artistes – the studios never stopped working! Denis Preston's production work was also very prolific.

Studio A and B had recording consoles with rotary faders. The main problem being that they used a combination of multiple studs and a metal wiper and needed frequent cleaning to prevent audible noise as the fader was rotated. IBC had an echo chamber down in the cellars, in close proximity to the boiler room. When there was a coal delivery the noise of the coal cascading down the coal chute was superimposed on the echo return – not good in the middle of a session! Studio A's recorder was a mono BTR2.

Studio B particularly suffered interference from Morse code, often preventing us from recording. It happened more than once when I was assisting on a session. When I first heard it over the monitors in B, it was pretty loud and seemed to swamp the equipment. I was told it was coming from the Chinese Embassy next door. I imagined they were transmitting enciphered Morse to China. Peter Harris: *"Interference from Morse code was not just in Studio B. The culprits were the Chinese Embassy on one side and the Polish Embassy on the other! We called in Post Office engineers to help. After a lot of trial and error, they produced a series of filters that were mainly successful, but sometimes the breakthrough was still too loud for us to continue."* The transmitter aerials were positioned on the roof of both buildings located either side of No.35; the transmitters were powerful, but fortunately transmissions were not continuous. Peter Harris again: *"Speaking of our close proximity to the Chinese Embassy, after the pianist Fou Ts'ong defected from China, we did a recording of him. He wanted a large grand piano that was too big to get upstairs. We therefore used the accounts office on the ground floor as a studio. In view of the possibility that people from the embassy might try to snatch him back, armed Special Branch police were stationed in reception."* There was certainly more than variety at IBC!

After a short period of time getting to know the place, all of us "trainees" were expected to set-up the studios for the engineers, mostly Studio A. Studio B was much smaller and used for small jazz bands, demos and DJ recordings for transmission on tape from Radio Luxembourg on frequency AM208. The edited tapes were sent over to the station. Radio Luxembourg at that time was a very popular radio station as there were no independent commercial radio stations in UK. As mentioned previously the BBC only had the Home Service and the Light Programme, hangovers from the Second World War.

When I had got my feet under the table, so to speak, I was eventually given more interesting things to do. I ultimately progressed to recording demos, in Studio B, for record companies to decide whether an artiste was worth recording and, if all was well, for the arranger to get a feel for what he was arranging, also to check the musical key for the voice range, and the general shape of the arrangement with key changes if required. My promotion to recording demos for record companies, in particular Philips Records with producer Johnny Franz, meant I could go from wearing a white coat to collar and tie, the expected standard at the time. It was a good start and while most demos in those days were piano and voice recordings more complex sessions was to come.

Studio B was equipped with an early all-valve (tube) Mono BTR-1 with tape oxide wound out – in other words the tape was threaded against inward facing erase/record/reply heads. One particular day at 2.00pm I was set up and ready to go for a Philips demo session. We did a lot of work for Philips, especially with Johnny Franz, who would bring artists to audition for possible later recordings. In walked a young woman whom I judged to be at about my own age, well dressed and smelling of a very strong perfume, almost overpowering. The pianist was there ready to play so I showed this young woman how far to be from the microphone and adjusted it for height. Johnny Franz stood on the studio floor listening to the performance, and for some reason that I could not fathom, on every one of these demo occasions – always in the afternoon – he always brought with him the lunchtime edition of the Evening Standard, reading it with one hand during playbacks, while playing pocket billiards with the other! We recorded two songs one of which I recall was "Burn My Candle" (at both ends). If the audition was successful this and the other song would be arranged by Wally Stott (Johnny Franz and Wally Stott always worked together) to be recorded later by an orchestra with the singer in Studio "A", recorded ensemble to a Mono BTR 2 sometimes running at 30 inches per second (76 cms). On this occasion the demo was passed and two songs duly recorded early in 1956, with Joe Meek as the engineer. I later understood that this lady came from Tiger Bay in Cardiff – the singer was Shirley Bassey and Burn My Candle (at both ends) was not a hit, although it was Bassey's first single. The BBC refused to play it because of its sexual connotations – still a very conservative broadcaster even then; Reith certainly left his mark. Stormy Weather was on the "B" side, and the single was released on a Philips 78 rpm shellac record.

On another session, I was working on an afternoon session with Joe Meek and the singer Ann Shelton, with Johnny Franz producing. We were recording "Lay Down Your Arms". It was decided by Franz that the song required the sound of marching feet between the choruses, which he was going to ask Joe to dub in. Joe came up with the idea to do it live as a performance. Franz left these effects matters to Joe, and I was told to find an empty box and fill it with stones – just like that! I used a BBC AXTB ribbon microphone box. Fortunately, I found in the street outside the studio that workmen had left a pile of white road gravel after some road repair or other. I filled the box, took it back to the studio, stood behind Anne who was singing into a Neumann U47, and duly shook the box at the appropriate times in the song. At the end of several takes I was covered in white stone dust – so was Anne, all over her posh frock! The song was a big chart hit for Anne Shelton, and went to #1 in 1956 in the UK, and stayed in the charts for thirteen weeks, and #59 in the US. As was usual then the vocal and orchestra were all recorded together – with performance edits where necessary by Joe. The song typically took about an hour to record during a three-hour recording session. In that era, all the recordings were in mono, mostly at 30 ips, 76 cms, and usually accomplished in just three or four takes, certainly no more, with perhaps some editing between takes if necessary for vocal or band mistakes. This was a rarity unless there were copying mistakes. It was the norm for artists to sing live with the orchestra, and, in my opinion, resulting in a good feel performance because of the interaction (the importance of a good studio set-up) between singer/s and the band. There was no selective foldback, therefore no headphones used by the artist/s.

Stott and Franz

It is worth mentioning here the working relationship Wally Stott had with Johnny Franz. In 1953, Stott became the musical director for a new British label, Philips Records. In those days, it was a relatively small label compared to HMV and Decca and did not have many staff. Wally Stott's job was to assist Johnny Franz in selecting material for the artists and then to arrange and conduct the recording sessions, which Johnny produced. At that time many of the British artists were covering American hit songs. Wally was determined to make the records as distinctive as possible. Frankie Vaughan's individuality was most effective in *Green Door* (1956) recorded by Joe at Conway Hall.

I was on that session with Joe, as his assistant, and what a powerful sound he created recording on the BTR2 at 30 inches per second! He really loaded the level on the tape, which would have made the techs apoplectic! The orchestration by Wally Stott strove to make *The Garden of Eden* (1957) more forceful than Frankie Vaughan's competitors – recorded by Joe again, in Conway Hall and banned by the conservative BBC. These recordings were made all the more powerful by Joe's technique of using closely placed microphones with copious amounts of "pumping compression/limiting" and reverb. In those days, many hit records were determined by the amount of plays on juke boxes in cafes, pubs and clubs, especially around the coffee houses in Soho and elsewhere. The audio level was set from inside the box and not available for the user to alter the volume. The technique therefore was to record and cut the disc "hot" with a reduced dynamic range, if possible only ± one dB, this is where Joe's limiting and compression techniques came to the fore.

With some of Joe's recordings the PPM (peak programme meter) needle hardly moved! With the compressed dynamic range and cut "hot", the record played louder than others in the juke box. So many of the records at that time were recorded by what Joe referred to as *"old fashioned engineers using antiquated microphone placement and who make whumpy sounds"* – a favourite expression of Joe's. Joe was referring to HMV studios. It seemed to me at that time, in the very early '50s, Decca engineers made better recordings than HMV. The culture at HMV appeared to be rather lodged in the past and the studios tended to be acoustically "loose" for pop, with no punch to speak of until much later in the fifties and early sixties. I asked Ken Townsend, former Operations Director of Abbey Road, about the pop engineers who worked there in the early HMV days: *"Laurie Bamber was the original pop engineer in the early '50s, who used a single microphone technique (Neumann M50). He was subsequently ousted by Peter Bown and Stuart Eltham, who did virtually everything and had loads of number one records between them. Then came Malcolm Addey who did all Cliff and the Shads. Peter Vince took over when Malcolm moved to the States. Norman Smith did all the early Beatles before going to Manchester Square. Geoff Emerick did a few (sic) albums, then Ken Scott."* Although George Martin was in-house producer at the time, Ken Townsend recorded *Why Don't we do it in the Road,* on the *White Album,* on his own with no producer! Ken also balanced a Shirley Bassey session when none of the other balance engineers was available for a variety of reasons! He also recorded *Theme One* (theme tune to BBC radio 1) at Westminster Cathedral with a small mobile unit (remote) written and produced by George Martin who played the cathedral's organ.

The looseness of the very early HMV recordings was also partly due to the lack of any really close-mic techniques being used in those days, although it has to be said that Studio 1 was always a superb space for recording classical music and in later years for film music scoring.

The same principle did not apply for Shirley Bassey and her IBC hits (Joe Meek) included *The Banana Boat Song* (1957) at #7, *As I Love You* (1959) at #2 and *Kiss Me Honey Honey Kiss Me* (1959) at #4. Robert Earl, who scored a hit with *I May Never Pass This Way Again* in 1958 at #14 said, "The combination of Wally Stott and Johnny Franz was a good amalgamation. They didn't believe in fade-out endings, so all those ballads end on big notes."

I wholeheartedly agree with that remark; I don't think the artistes would have enjoyed such huge hits if Joe had not been involved in the recordings. He excelled at innovation and experimentation, much to the horror of some of the techies, especially when he pushed the boundaries. In that era, little or no credit was given to the engineer's skills, and their creative use of the sparse technological tools they had at that time. They could literally make or break an artist through clever application of their skill on the one hand, or lousy recording techniques or "no ears" (lugs as we called them) on the other. Later, Denis Preston changed this lack of understanding of the engineers' role by always crediting them on his records. The number and scale of the hits was most certainly good for IBC, and for Joe too.

Studio B was a busy small studio and had grams (in the BBC parlance used at the time): two gramophone turntables running at 78 rpm, for playing 10″ and 12″ records respectively. The record sat on a "slip" mat and could be held with the turntable running. With the pick up on the disc and using pre-listen, I cued up the start of the sound then turned back half a turn and held it until the presenter had finished his segment – I followed the script – then span in the disc and opened up the main potentiometer to the console. I got very adept at spinning in records (on time) and from then on was put on every recorded show we did for Luxembourg. I was never late or missed playing in a cue when on the road with Joe – you would never hear the last of it if you did!

Chapter 3
Working with Joe Meek

A great deal has been written and said about Joe, with quite a mystique built up over the years – not all of it accurate, particularly when I reflect now about the many unkind things said about him by some members of staff and management at IBC. I worked with Joe for three years at IBC and a further eleven months at Lansdowne. I knew him very well. He was an enormous but volatile talent. I believe much of what has been written, and especially the film made about him, did not accurately portray the early years, certainly not in my experience of working with Joe at IBC and Lansdowne, before he left in November 1959. I feel I should at least attempt to redress the balance.

Joe Meek was originally employed at IBC as an assistant engineer. After he was promoted to sound balance engineer, with me mostly as his assistant (after Jimmy Lock), Joe and I eventually went "on the road" winter and summer, with a short break for a couple of months, to record the shows for Radio Luxembourg. Once the shows were in the can, all of us – cast, crew and producer John Whitney – went out to dinner. They were always good-humoured relaxing affairs, often lasting quite late into the night. On one occasion, John Whitney gave me a *very fast* lift back to London from Clacton in his brand new Jaguar convertible with the hood down – when I watched the first Bond film, *Dr No*, it reminded me of that time when I saw Bond driving an Aston Martin similarly fast. The music score for that film was written by John Barry, and recorded at CTS Bayswater, a studio I came to own much later at Wembley with Johnny Pearson. Small world!

We travelled on Fridays and recorded the show on Saturdays. I took over the position to assist Joe from Jimmy Lock and it involved travelling all over Britain during the summer recording When You're Smiling at many of the Butlins holiday camps, such as Clacton, Skegness (the first camp to be opened in 1936) Paignton, Ayr, Filey, Rhyl, Scarborough and Pwllheli.

We ate lunch in the dining hall with hundreds of other holidaymakers and were put up at hotels in the area. In the winter, Joe and I went to many music halls recording another show for Luxembourg – *People are Funny* – again staying in good hotels such as the Grand in Blackpool. Sadly many of these halls are now demolished, such as Chiswick Empire, Collins Music Hall (only the facade remains today), Metropolitan Edgware Road and Greens Playhouse Glasgow to name but a few. We also recorded in and around London, and in some other venues, a show called *Shilling a Second* – the original recording engineer for this show being Peter Harris. I remember very well we travelled from Blackpool to Belfast in a very old twin engine boneshaker of an aircraft chartered by Hector Ross Radio Productions with no cabin heating – "Broken!" explained the Captain – truly awful and very cold! At lunchtime we went to a pub for a pint of the black stuff, and although the pub had to close at 3.00pm, we weren't kicked out as the doors were locked and the drinkers had other pints lined up – mostly Guinness – all along the bar, bought before closing time, and continued drinking until the pub opened again.

"Be Jasus, the Garda can do nothing they're just finishing their drinks they are paid for," said the landlord. In the evening, Joe went out alone and I went out with the air crew and ended up at 3.00 in the morning drinking tea and Paddy Irish whisky mixed in some God forsaken bar – never again! I had one hell of a hangover in the morning when setting up at 8.30am! Joe turned up late with someone he described as his "friend" in tow. "Friend" didn't hang around long.

Before travelling to record a show, on Friday afternoons, I would check out and prepare the equipment, and make sure we had enough cable reels, and that any faults in the cable connectors from the week before were fixed by the workshop. The old large type of canon connector was notoriously prone to failure in that the solder joints were easily pulled apart in the connector, the connector cable clamp not being very well designed. On strip-out, I wound the cables back onto a cable drum as fast as I could. Sometimes the cable connector would snag on something or other and pull the connector solder joints apart. The workshop was not amused, and thought Joe and I should take more care of the equipment. I learnt my lesson there – it gave the workshop grief and earned a moan to Joe and me. Later in my career, I became very strict with my engineers in not pulling cables when they got jammed.

I had to also check we had enough microphones. These were made by Standard Telephone and Cable (STC, now Coles), and were all dynamic types, Ball and Biscuit (omnidirectional) type 4021, Cardiod type 4033a, Ribbon (bi-directional) type 4038 and Pencil type omnidirectional 4037 for hand-held on the stage. There were two Vortexion four channel mixers with microphone amplifiers, linked to make eight channel inputs with a standard BBC PPM (peak programme meter on one mixer. For public address (PA), where possible we tried to get a feed into the theatre PA system. This was not always possible as many of the theatre systems in those days were past their sell-by-date or 100-volt line systems, often not working, with dusty old public address amplifiers with valves covered in years of dust and dirt (it was a wonder some worked at all) that were positively dangerous devices with high voltages present. When we had problems like that it was my unfortunate lot to rig our own flying-saucer type PA speakers to hang from the ceiling. No mean task. It meant crawling around the ceiling space in dust and dirt – no Health and Safety then, or any safety officer watching over you, and no hard hats – so we avoided doing it when we could. When I look back it was a very risky business, thank goodness I didn't have to do that too often. I recall Greens Theatre in Glasgow was the worst for rigging. Their PA system was out of the dark ages and the PA amplifier was covered in decades of dust and dirt. It had certainly seen better days. When I enquired if we could hook into it, to save running our cable PA system, the reply was, "It doesn't work mate! Not worked for years." It seemed to me at the time that many of the older theatres did not have any spare cash around to maintain the infrastructure. In today's world there would be a rigging crew, not just Joe and myself.

We had two Vortexion tape recorders running at 7.5 inches per second – one for recording, the other for music play-in and change over for continuous recording. For *People are Funny* we used the second machine for playback of Bob Danvers Walker's street interviews, recorded at 7.5 inches/second on the EMI L12 portable. I had to be pretty nifty to remove Bob's recording from the EMI portable, load the second machine and cue it up for audience playback and, if Bob was late returning and had to get on stage almost immediately, it was a close-run thing.

The regular routine involved all the kit being picked up on a Friday afternoon and delivered on the Saturday by lorry to whatever venue we were going to. We had to be ready for our 8 o'clock in the morning starts on the Sunday, and be set up ready by lunchtime or very early afternoon for the rehearsals.

Joe and I met on the Saturday morning, at Kings Cross, Euston or St Pancras station, travelling by steam train to a northern theatre or a Butlins holiday camp in the summer. We would travel from Paddington station if going westwards and Joe would sit and write his songs on the journey. On one trip, on the way to Blackpool, I remember the makings of a song – the lyrics – of *Put a Ring on Her*

Finger. After lots of scrawling, crossings out and partially finished notes it was eventually recorded by Mike Preston, Les Paul and Mary Ford. Joe couldn't write music. I don't think it came easy for Joe to write lyrics, as he was more of a technical guy and would sometimes stop writing and reminisce about his days in the RAF on radar watches, just two people, and how they used to eat their food – "Dinner" Joe called it whether lunch or evening meal – using the large capacitors of the installation as a table! He would also talk about his days as a TV repair engineer, based in a workshop of whichever firm he was working for, either Currys or Midlands Electricity Board then a company called Broadmeads. In an odd way, I think he missed his rural town past. Joe only occasionally spoke to me about his past in Newent or his family.

On the lengthy steam train journeys we would eat lunch in the restaurant car, with silver service and white-jacketed waiters wearing white gloves with immaculate white linen tablecloths and napkins. Whatever happened to those days, when travelling by train and eating on board was such utter pleasure, albeit with longer journey times? These meals were on expenses of course! Ten bob (£12.25 today) or fifteen bob (£18.37) for a very good 3-course meal. On one occasion journeying to Blackpool, we left to go to the restaurant car: 1st class for lunch. At the entrance to the coach, Joe stopped. I said,

"Joe why have you stopped?"

"I'm looking for a waiter with a big nose so we sit at the table he serves," He answered vaguely.

"What? Why?" I replied.

"If he's got a big nose then he'll have big one!" was his reply.

That was the first time I heard Joe talk publically about his sexuality. It should be remembered that in those days being a homosexual was a criminal offence, punishable by a harsh prison sentence. The word "gay" was not used in that context. Society deemed that these matters were best kept quiet. I got on well with Joe and when I reflect back, he was an exceptionally talented engineer. He was greatly underestimated by the others at IBC as he invented very creative, and pioneering, recording techniques that would upset the more traditional engineers. Denis Preston recognised his enormous talent and requested that Joe record all his sessions. That put a few noses out of joint, as Denis was a major client of IBC.

After we had eaten dinner in whatever hotel we were staying in, Joe would say, "Well I'm off now, see you in the morning."

He would arrive at the theatre, usually late, with a person of whom he said, "Let me introduce you to my friend." The "friend" didn't stay long for long. Joe wanted to let them know what he was doing, and took great pride in showing the "friend" around but that was it: nothing more said at all, no references to his "friend" – forgotten! I realised then why he went out on his own in Belfast and regularly after dinner at other venues.

Whether he had a conscience or not I never found out, and made it my business not to ask. He had his permanent boyfriend back in London; a chef by the name of Lionel. Every venue we went to it was the same routine, unless he got unlucky. In the evenings after dinner and once Joe had "gone out", I would take the L12 portable recorder and capture traffic street sounds and other general street effects, passers by walking on hard surfaces and the way the sound changed when walking past a large set-back shop entrance and so on.

Following a good breakfast, Joe and I would travel back to the studio on a Monday morning and on Tuesday I transferred the 7.5 inches per second (19cms/sec) tapes to a BTR2 running at 30ips (76 cms/sec) in what was known as the assembly room on the mezzanine floor with a glass observation window overlooking Studio A floor. There was some other gear in this room, in particular a very fierce limiter that could be made to "pump" when driven hard. It had a Magic Green Eye as the indicator of the desired limiting level, and as more input signal level was applied, the more the eye would close up.

Joe usually set it to maximum on his sessions if using it live. We also had what was referred to as a "curve bender", which Joe called the "cooker" – in today's parlance, multi frequency equalisation. Joe used the limiter to great effect and this can be heard on Humphrey Lyttleton's *Bad Penny Blues*, along with the "cooker" being used to exaggerate the top and bass.

Humphrey Lyttleton's *Bad Penny Blues* was recorded in mono at 30" per second (76cms), and Joe had used close-mic techniques on every instrument, in addition to compressing the music like a madman. The producer on the session was Denis Preston, whose record contract was with EMI Parlophone, so we had to cut the record (master lacquer) at Abbey Road, which in those days was even worse than IBC when it came to the "men in white coats" syndrome. If the record was not considered technically perfect, in their opinion, they wouldn't cut it. When we sent the master round, they just didn't know what to do with it. They returned it, saying it was too distorted to cut. Joe started ranting and raving: "That's not distortion. That's commercial. Rotten pigs. Send it back now!"

Eventually, Abbey Road did cut it, and it was a hit in July 1956 at #17. Denis Preston, now and again, would create alternative names for places or people. Denis referred to Abbey Road, at that time, as "Shabby Road"! On one occasion I was sent to Abbey Road to deliver a master tape and while there decided to have a cup of tea in their canteen. What has always stuck in my mind was the sugar bowl on the counter with a spoon attached to the counter with a piece of string. I guess they thought the spoon would get nicked!

Much has been documented about *"Bad Penny Blues"*. I was present on that session as Joe's assistant. Joe and Denis always worked closely together. Joe had a concept of sounds that certainly Denis as producer didn't always suggest, but they worked very much as a team and Denis let Joe "use his noddle" (sic). For example it was Denis who suggested the close up drum sound, and he wanted more "fizz", or heavy top-end equalisation. Joe close-miked the bass end of the piano, to get the close-up left-hand boogie-woogie sound, close-miked the drums and tightly miked the acoustic bass. Humph's trumpet was close-miked to a U47 – and I mean close – he used a straight mute with plunger giving that unique sound. The recorded master take, with heavy piano bass and acoustic bass, plenty of reverb, was played back to the band including Humph – and there were no complaints!

It was a different concept with elevated drums, piano and bass with a good audio perspective and a very good "original" recording, captured with Joe's disregard for tape level, with the machine level meter showing well into the red. After the session ended and the band had departed, Joe worked his post-production "magic" with more equalisation getting the "cooker" and the limiter pumping. When Humph eventually (I believe he went on holiday) heard the final released record he hated it!! Hate or not, it was the first crossover Jazz hit. It's still being aired today, sixty years later, and can be bought on iTunes! Denis used to say, *"Jazz is only jazz if it doesn't sell, if it sells it's called commercial."*

Almost everything Denis did with Joe was a hit. *Bad Penny Blues* is but one example. Joe's concept of the sound under primitive conditions was, in Preston's view, ten years ahead of his time. Denis Preston: *"Any idiot, and I say this surprisingly, can use the most complex board but you can have someone without any tech knowledge, Joe was technical, he had lugs (sic), and he also had a heart."*

On a Wednesday following our weekend's Luxembourg show recording, under the supervision of John Whitney, I would also edit the show in the assembly room. This meant cutting the tape with demagnetised scissors, or non-magnetic brass ones, at an angle judged by sight to be 45 degrees. No edit blocks or electronic waveform then to be of help, you had to get the cuts correct every time and stick the joint together with editing tape that was slightly less than a ¼" wide. Audio editing consisted of scrubbing the tape back and forth by hand against the replay head, listening for the place you wished to edit in very slow motion, marking the cut, then finding the place for the second cut, marking both places with usually yellow or white Chinagraph pencil, removing the piece of audio you don't want and splicing the tape back together again. It was an acquired listening and scrubbing skill that took time to perfect.

The AGFA recording tape we used on the studio BTR 2s, as well as on an extremely heavy transportable machine with separate transport and valve-electronics boxes, was 3,250 feet in length of tape, wound on a metal centre only called pancakes, which then sat on a platter on the machine. It was a fairly abrasive recording tape and would wear heads down pretty quickly, especially if spooling and occasionally pushing the tape against the heads to check where you were to find take after take. One had to be very careful when taking the tape out of its box. If the centre dropped out when removing it from its tape box or the spooled recorded master tape was not wound tightly enough due to the machine fast forward or rewind needing adjustment, one had a major disaster with tape everywhere! It would then mean I had to rewind the whole lot back from the floor, making sure I got the tape the correct way up so as not to play backwards – a long and laborious job. Fortunately, on the back of the AGFA magnetic recording tape, AGFA printed the tape type, and that made the job much easier. It happened to me once when later editing a Laurie Johnson master with Denis Preston, I had a difficult edit to make cutting on a drum beat in a fast tempo piece, the drummer was Phil Seaman who was a great player but never played the same thing twice with the odd fill here and there, not written on the part, I made the edit but it was very slightly out in timing. I had to cut back in (with scissors) a small piece of tape about two eights of an inch in length. To redo the edit I took the tape off the machine and the whole lot went on the assembly room floor.

It took me a considerable amount of time to wind the tape back onto the machine and save the master. Denis found it amusing. But you only did that once!

My only day off during the busy show period was a Thursday. I was paid fifteen shillings (£18.37 in today's money) a show, all expenses paid, in addition to my princely salary of 5 guineas a week (£130.92 per week today). Things were much cheaper then and we were quite happy with our lot and, although I was living at home, my parents refused a contribution toward household expenses. Joe took care of all spending money and was meticulous with obtaining all receipts; I believe he was given a spending float courtesy of Hector Ross Radio Productions, who produced the shows for Radio Luxembourg. Known as Two O Eight (208) the long-wave frequency transmitted in metres. We didn't record shows all year round. We did some in the winter months then had a break until we went on the summer show at Butlins Holiday camps.

Homosexuality in the 1950s and 60s – the Turing Parallel

As background to this aspect of Joe's personality, it is worth mentioning Alan Turing here. Dr Turing worked at Bletchley Park during the war, and is now recognised as the father of computing. Together with Tommy Flowers from the GPO (General Post Office) at Dollis Hill, he built the first viable computer, Colossus, to crack the German enigma codes. Turing was homosexual; the law then criminalised sexual acts between men, even in private. In 1952, Turing was convicted for gross indecency to which he confessed. He was given a choice, imprisonment or treatment – "a cure" – he chose the latter. In lieu of a prison sentence, Turin opted for a hormone called oestrogen, delivered by injection to reduce his libido – chemical castration – reversible when treatment stopped. What an absolutely disgraceful way to treat such a talented man who did so much for his country. It is said he committed suicide on June 7th 1954. He died of cyanide poisoning, a half-eaten apple beside his bed. The apple was never tested forensically. This is possible conjecture; at home he conducted experiments electroplating spoons with gold, a process that requires potassium cyanide, and it is very possible he accidentally inhaled the cyanide vapours from the liquid, which would pervade the room.

Homosexuality was finally recognised by law in 1967. It took the Coalition Government of the day 61 years to apologise, and to recognise Turin's huge contribution to Britain's war effort. A pardon was granted in December 2013 under the Royal Prerogative of Mercy, after a request by Justice Minister Chris Grayling, who said, *"In my maiden speech here in this House, I spoke of Alan Turing, the code-breaker who lived in my constituency, who did more than almost any other single person to win the war, and who was persecuted for his sexuality by the country he helped save. I am delighted that he has finally received a posthumous royal pardon."*

You may wonder why I mention this and what it has got to do with Joe. There are definite parallels with Turing's situation given Joe's exceptional talent in his own field. Stagg didn't think so, "Not a great engineer", was his diagnosis. He was lucky not to be caught, especially on our trips when he would openly hang around known meeting places for homosexual men – a huge risk. He in turn made a colossal contribution to the recording world in the UK, creating techniques with the primitive equipment available then many of which are still in use today, especially close-miking, clever use of reverberation and limiting/compression. If Joe had been caught and convicted, a prison sentence would have been inevitable and our recording industry would have been a much poorer place. Later when Joe was at Holloway Road he was caught in a gentleman's convenience "soliciting for an immoral purpose". He was found guilty of smiling at an old man. Joe was horrified, saying, "I don't go smiling at old men! It's young chickens I'm after!" Police officers from the "fairy squad" were keeping watch on the loo in Madras Place in Holloway Road, from Islington Library. Joe went before the magistrates and was fined £15! (Source: John Repsch) Tragically, Joe had died by his own hand by the time homosexuality was legal.

Knowing Joe as well as I did, my perspective about his death was that he could be very temperamental in getting his own way and often lost his temper quickly especially when taking, as he did at Lansdowne, Preludin tablets in quantity. A slimming pill, or upper, called phenmetrazine, it is described as the most euphoric and pro-sexual of the stimulants. He needed it to stay awake, as he was frequently up all night working on his recordings especially at Lansdowne. I firmly believe that the drug caused him to be restless, panicky and fatigued. Joe could also be very paranoid. He often threw things around at Lansdowne, and I kept well clear. He used to give me two shillings and sixpence, whereby I had to run an errand to the local chemist to buy the pills over the counter for him. When he lost his temper, as he did when taking the drug, afterwards he reflected on what he had said or done and was regretful. Joe, another talented person like Turing, Joe highly committed to the recording industry, sadly committed suicide by first shooting his landlady Violet Shenton, then turning the shotgun on himself on 3rd February 1967. Knowing Joe over the years as well as I did, I suspect he must have again lost his temper with Mrs Shenton (see John Repsch: *The Legendary Joe Meek* account, page 307 onward), taken revenge with the shotgun, regretted it and finally shot himself. Naturally, I was deeply saddened when I heard about Joe taking his own life on that fateful day. I owed him much in helping me in my career, and fortunately his pioneering legacy lives on.

Denis Preston on Joe's death: *"He was probably one of the most original and creative recording engineers in the business and I think paved a path for a lot of things that were outrageous in his day. He was*

*a pioneer, a bit of a prophet and a marvellous nut, and I can say no
more than God rest his soul in everlasting noise."*

Joe's main theme was to write songs on trains and experiment in
the studio with his own sounds, much to Stagg's dismay and dislike.
Typically, three or four songs completely absorbed him and we
would arrive at some cold blustery place up north and he would have
written the songs on the journey. This was Joe, completely absorbed
in his songs, his music and his work. A complete natural talent for
what he was doing. He followed through from the technical side, to
the musical side and the production side – what a great experience
for me working with him all those years. He aimed to get the direct
sound, not that old roomy sound. Joe got a result for the sound that
felt right and sounded right for the job he was doing. People threw
their hands up in horror and said, "That will never work!" Well, to
use Joe's words, "Rotten Pigs!"

I do not seek to denigrate the other engineers at IBC, as they
were indeed very good engineers, some steeped in the old ways of
recording, and having said that, Joe was ahead of the curve in what
he did, taking technical chances on most occasions – pushing the
technical boundaries with the now seemingly archaic equipment
at our disposal. IBC was a superb training environment; many
engineers who worked or trained there later went on to be heads of
other successful studios in London and producers in their own right.

I just accepted Joe's sexuality for what it was. I guess that was
one of the reasons we got on so well because he was occasionally
humiliated at IBC by his colleagues, they called him names behind
his back, and at staff meetings particularly he was every now and
then reduced to tears. Allen Stagg could be particularly snide in his
comments at those meetings. Most cruel, and something he didn't
deserve – jealousy I guess because Joe was very much in demand by
IBC clients especially by Denis Preston who always requested Joe
to record his sessions. Denis recognised Joe's talent and thought the

other engineers at IBC were not up to Joe's creative standards – it put some other engineers' noses out of joint. Other engineers were sometimes given the job even though Joe was available, and that was abhorrent to Denis who complained to Allen Stagg. Denis referred to Stagg as "Edgar G. Stagg" (typical Denis) and Joe was upset by not being given the jobs, despite being requested by Denis or another client. Stagg went out of his way to ensure Joe was not on some of Preston's sessions – most unfair. Joe's creativity could strike at any time, and as an example, I recall a lunchtime visit to a wireless surplus shop called Smiths, in Edgware Road. While rummaging through some boxes of parts, Joe picked out an enclosed clockwork timer or time switch. He set it going and held it up to his ear. As it ticked away, he said, "I like the sound, and its rhythm. I can use this on a record." He was always on the lookout for ideas that would bring originality to his work.

Joe may have been highly temperamental, but he had a good sense of humour when not provoked by others. I think there were some at IBC who couldn't read Joe too well, they didn't understand his talent. There was much prejudice against him, with people looking for other meanings because of his sexuality. If he didn't like one of the assistants he was assigned to work with, he refused to work with them (they nearly always complained to Stagg) and, as for producers that made stupid comments or didn't understand what was going on, when it happened, well... One thing for sure, prejudice was conditioned because Joe couldn't always get his own way and would be antagonised by others. A good example was the "cooker" – Joe's terminology for the multi-frequency band equaliser and a "limiter" in the Assembly Room. If you had a session and wanted to use either of those two pieces of kit it would have to be shared between the two studios and booked, such was the shortage of outboard equipment – zilch! Engineers would plug into the outboard kit by tie lines from whichever studio they were working in. When Joe wanted to use the "cooker" or "limiter" someone else would be using it and Joe would say "well I've booked it", and he had! I can vouch for that, as I frequently went to Bookings to request it for Joe, or he would go himself. I clearly remember scenes like that. Some of the engineers would go out of their way to be obstructive over the equipment usage and thus prevent Joe from using it. I thought it was not imperative, in some instances that they needed to use it at all. Joe would then create a scene because he didn't have the pieces of equipment, the tools of the job that he had used so successfully before on many hits. He would use his favourite words "Those rotten pigs". The limiter had a green magic (the same as the tuning eye in old valve radio receivers) eye that closed up more and more, the harder you drove the device. When Joe was using the limiter, I would have to go from the studio up to the assembly room and let him know how closed the eye was. Joe mostly used the eye closed, or with the green quadrants overlapping each other, and it really did pump heavily – e.g. *Bad Penny Blues*. This, combined with heavy equalisation, was Joe's audio signature and became more evident in his later years. *Telstar,* recorded at Holloway Road, is a particularly good example.

What I can say about Joe was that his personal life never ever interfered with his professional life or with his creative life; he was always absolutely focused on the project in hand.

Joe and I never had cross words. We understood each other perfectly, and had a good working relationship and a good laugh. He frequently combed his hair to keep his quiff in place and I'd say, "It looks good to me Joe." On his personal appearance he always, without fail, wore his dark Slim Jim tie and jacket. The exception to the rule was on sessions in a hot control room, and it WAS hot especially in summer, as IBC had no air conditioning. When there were band sessions in the evening, neighbours living behind the studio building often complained about the noise (nothing new here), particularly if the musicians asked for the windows to be opened on a hot night. Those large fans came into operation again, installed at the front of the studio.

With Joe at Conway Hall

Located in London's Red Lion Square, Conway Hall had excellent acoustics and a fairly large stage, where on occasions *Shilling a Second* – billed as *CWS Margarine Show* (Co-Op Wholesale Society), "Britain's Brightest and Best Radio Quiz" – was recorded for Radio Luxembourg. The show featured announcer Patrick Allen and Canadian Compere Gerry Wilmot, with Joe as the engineer and me assisting.

The hall had sufficient seating to have capacity for the show's audience and could comfortably accommodate well over a sixty-piece orchestra. The choice of microphones for the different orchestral instruments was critical. Some of the artists we recorded there were Peter Knight Orchestra, Edmund Hockeridge, and Harry Secombe, all involving large orchestras. Other artists included Cyril Stapleton Orchestra, and the Eric Delany band – he had masses of percussion, the like of which I had never seen before. The big band sound in that acoustic sounded tremendous as Joe made the hall work for him, and he knew exactly where to place the guys so they could hear each other well but still get the results Joe wanted. No headphone foldback then and NO equalisation on the console. The careful choice of mics was vital. Those guys had to see and hear each other to perform, and Joe organised the sessions accordingly, such was his sensitivity to the artistes. He had good ideas and usually got his way, if not there could be a scene but thankfully that did not happen often when the team comprised Joe, myself, and the producer. The producers who respected Joe, and there were many, let him get on with it. As an example, we built a large canopy over Eric Delany's huge percussion and drum kit with blanket-like material, and this tightened the percussion sound for recording, and left the band in the open hall to achieve a good big band sound with "air" around the instruments.

We didn't want some of the small Latin percussion rattling around the hall – we needed the separation.

Later in my career, I used many of those simple but effective techniques, particularly in the '60s, on sessions in Cologne. The one thing that got up Joe's nose was when producers asked him how he got the sounds that he did. That was red rag to a bull and the usual phrase came trotting out "Those rotten pigs trying to steal my secrets." He sometimes wouldn't even tell me what he was up to but I worked it out eventually. He was way ahead of his time, getting the most out of the somewhat limited equipment we had at our disposal. But he was also paranoid about keeping his secrets – his was a hard won success.

I was Joe's assistant on many record recordings at Conway Hall, although the small control room was permanent for IBC, we had to bring in the portable BTR2 recorder and microphones transported in the rickety old IBC van, the same van we went in to the Eisteddfod at Aberdare in Wales. For once, at Conway Hall Joe didn't fiddle with the recording curve equalisation! The workshop guys had set it up and Joe kept his screwdriver in his pocket. It was a marvellous hall to record in, with a wonderful natural acoustic space – superb for orchestral strings and orchestra.

All sessions were recorded at 30ips in mono; stereo had not yet come of age at IBC in 1955/1956. The orchestral balance had be right more or less first time; no time to faff around, choose the right microphones for the orchestra sections, get the orchestral layout correct, make sure the mikes were properly identified, i.e. the correct mic input to the chosen input channel – "scratch around" we call it. The conductor's stick goes up, the band play, you open up the faders and create the sound, with no more than two run-throughs, if there were no score parts copying mistakes, which did happen from-time-to-time. Then you start to record, announcing over the talkback "Tape's Rolling, take One", maybe a master recording in no more than three/four takes, edits as required: that's it job done! Many engineers today (and producers) would panic at the thought, no multi-track to get them off the hook! If a well-rehearsed vocalist was involved, then he/she sang live with the orchestra. Today, some producers would have take after take and kill the spontaneity of the playing, mostly because they couldn't decide on what was good or not, whereas the band and an experienced engineer would know decisively whether the take was good or not! Later in my career, I worked with quite a number of producers like that, some even tone deaf – God knows how they got the job! They lacked confidence in their own ability, and would sometimes panic: "*What?* We have run out of tracks?" I recall us working with Harry Secombe at Conway Hall, recording classical works, arias etc, all live with the orchestra. If he made a mistake, which sometimes occurred, he would stop singing and blow one of his huge raspberries Goon Show style; we all collapsed into laughter. To balance a large orchestra in that hall was relatively easy because of the good natural acoustic. I learnt much about sound balance engineering in a live acoustic.

IBC also had GPO tie-lines back to the studios at Portland Place, so if we needed for some reason or other needed to add additional echo (reverberation), we sent a signal to the IBC echo chamber and returned it back again. The restricted line bandwidth was not noticeable for echo. On another occasion, we were recording Cyril Stapleton Orchestra with the main orchestra in Conway Hall and the brass in Studio A at IBC, while recording onto machines in the Assembly Room. The first time this had ever been done. Allen Stagg was at balancing at Conway Hall and Eric Tomlinson balancing in Studio A.

When we were recording at Conway Hall we were fed by Lionel, Joe's long-term boyfriend, for absolutely nothing – he ran a cafe not a stone's throw from the hall. That pleased Joe. Joe knew I respected his ability and where I personally stood. He respected me for never going on about his sexuality, as others did at IBC. To some members it was utterly abhorrent and they did not hide their distaste.

Joe was an enormous trendsetter for the recording industry and in many quarters held in high esteem, even today, for his recording skills and foresight. His innovative approach to recording, with careful microphone placement, often resulted in some saying you can't do that – referring to close miking – as it'll overload the mike input. Sometimes it did, but the problem could be resolved by a passive attenuator on the mic input amplifier, and Joe used the technique to great effect. One has to remember that recording consoles in those days did not have the headroom found in today's technology. Joe was in immense demand by IBC clients, and invariably and regularly engineered hits. We used to refer to it as Painting Pictures in Sound, for that was what it was. You start with a blank canvas and "paint" the sound with colour and depth. Difficult to achieve in mono, but so much better later in stereo.

When I worked with Joe, particularly in Studio B where he recorded small jazz bands for Denis Preston, he would always adjust the internal recording curve equalisers on the machine after it had been carefully aligned by the workshop. On one occasion I asked why he did it, his reply was "To get more top". It caused the workshop much grief, and other engineers too, who left well alone – in any case, the CCIR/IEC recording curve was the UK/European standard and not to be messed with. A calibrated test tape was used to line up the tape machine. There were no equalisers in the recording console channels, and IBC made two portable ones in boxes, in the hope that Joe would no longer fiddle with the machines' record EQ. They could be patched into the recording chain.

Joe saw a new pioneering way forward with microphone techniques and is now generally recognised as the father of close microphone placement techniques – signal processing notwithstanding – in a limited way because of the primitive technology used in many of today's popular music recordings. Some may disagree. I do not. I was there when Joe experimented, much to the annoyance of others at IBC.

You only have to listen to some pre-1950 records to hear how bad they were, with the booming room acoustics. I called it "off the wall recording" (also bad studio acoustic design), because microphones were at a distance from the players. There were engineers at IBC, who by their own admission "did things (recordings) the standard way".

Allen Stagg was more of a traditional classical engineer, not into pop or jazz. Keith Grant, another highly talented engineer joined IBC as a trainee, after I left for my enforced two-year gap for National Service. On Keith Grant, Peter Harris again: "*Keith Grant was another engineer who had his own techniques. One of his specialities was to disregard the need for matching 600 Ohm outputs and inputs, and hung everything he felt like hanging together. Nevertheless, he got results.*"

Joe regularly experimented with close-mike placement. I particularly remember one session at Conway Hall in the mid 50s with Eric Delany band – Delany was a drummer and percussionist. Joe placed two STC 4037 (pencil) microphones pointing up inside the bongos, one for each drum – the close sound was incredible – you could actually hear the finger movements on the skin of the drums. When Joe played back the result in studio A control room, the others at IBC were amazed, and I certainly was!

And so for Joe who called some of his colleagues at IBC "Those rotten pigs", his favourite term for anybody who did not understand about him or his recording techniques. Peter Harris recalls: "*I remember when Joe threw a fit because he wasn't allowed to record a certain artist that he believed he had a right to work with. He asked my opinion and I started to suggest that it was down to the management to decide these things. He flew into a super tantrum and accused me of being the same as all the others. I believe that he took his revenge when I found that all my faders had been sabotaged when I unpacked my equipment for the 'Shilling a Second" recording on the following Sunday. Luckily, the venue was the Croydon Empire, so I had time to return to the studio and effect repairs."*

Other sessions at Conway Hall were equally fruitful when Joe could exercise his talents without the constraints of the studio or the innate criticism he had to endure back at base: management criticism of the time he spent overnight making client masters sound good, was the time being logged or was it for Joe?

Bending the Curve

On occasions back at base, Joe became paranoid *again* about people stealing his secrets, and this tended to make him a loner. He mostly did his own thing on sessions and did not want others to know how he achieved the sound, especially tweaking the CCIR 1953 Standard (Comité Consultatif International des Radiocommunications) recorder equalisation or in today's parlance recording curve (International Electrotechnical Commission) IEC/DIN 1968 current standard. For all intents and purposes, CCIR and IEC/DIN are one and the same, an updating of terminology. In the USA, the standard is the National Associations of Radio and Television Broadcasters, now shortened to NAB, recording curve. All curves, known as standard flux curves, are expressed in nWb/m (nanoWebers/metre). These standard flux curves are applicable to professional studio recording speeds of 7.5 & 15ips using an international standard calibration tape to align the recorder – IEC, NAB, AES. The NAB standard test tape flux levels are lower than the European ones. There is another curve for 30ips (76cms) recording, (adopted as an international standard at that speed) called the AES curve – Audio Engineering Society – AES announced the adoption of this curve in 1951, and it was used by the Ampex Corporation in its recorders. As can be seen, it is a complex subject and outside the scope of this book. As a matter of historical interest, the NAB curve was designed by Frank Lennert of the Ampex Corporation for the Ampex tape machines and was approved as an NAB Standard in 1953.

On the BTRs, Joe would adjust the curve to get more top end, much to the annoyance of the guys in the workshop, who had to realign the machines again and again! Sometimes, Joe would note where the potentiometers settings inside the machine were originally set, and then put them back after the session – but not always accurately. Joe not only fiddled with the recording curve adjustments, more top added, but he also fiddled with the playback curve, and so it could be a nightmare for the other engineers if Joe had been working before them, as they never knew what he had been up to – more work for the workshop. I told Joe he couldn't do that as it must be set correctly, and his usual answer, yet again, "Those rotten pigs don't understand". In other words, he was using the machine recording curve as an equaliser. Another of Joe's techniques was to pile on the recording level so as to overload the tape, to achieve the tape compression and distortion that went with it – tape saturation. As the signal level increases to tape, the result is distortion and compression, which behaves in a non-linear way – that is, unevenly with regard to signal level, frequency and dynamic range, critical at high frequencies. Even more so with Joe increasing the recording curve by lifting the high frequencies on recording, because magnetic tapes have reduced headroom at high frequencies. The end result is an overloaded signal that displays natural compression and limiting characteristics. Not an unpleasant sound if used carefully. With the advent of Digital Audio Work Stations (DAWS) today, plug-ins are available to emulate tape saturation and other analogue effects from past years.

The recording tape used was AGFA, which could accept a very high recording level and tape saturation. It could become a technical nightmare for the maintenance staff as one never knew which machines Joe had been fiddling with. He did care really but was intent on making his own "audio signature", as I call it today, or as he called it, the RGM (Robert George Meek) sound. It should be remembered that in that era there was no multitrack, it was recording direct in mono to ¼" tape, therefore the creative work had to be made live, with the band and singers all together in the studio and with very little equalisation or sometimes none available, so the choice of microphones to use on different instruments was paramount. Joe overcame the multi-track problem by creating composite masters, bearing in mind there were no stereo recorders at IBC in those days. If an artist couldn't cut it "all up" in the original studio recording, we recorded the orchestra track at 30ips (usually the norm), played it back and added the vocal whilst recording onto a second BTR at 30ips – very little loss if the machine was correctly lined up! The analogue sound at 30ips was superb, and it was almost impossible to hear any degradation of the recorded signal from the input signal.

At high recording levels, the ¼" tape noise floor was almost insignificant and we could therefore create several composite overdubs that way with negligible loss of sound quality. This process allowed Joe to further compress/limit, equalise or add further echo chamber reverb on each pass if necessary. As his assistant I was allowed with him but no others were around – it's a secret you see! We did this on most major artists' recordings if the artistic creativity required it. One producer, Michael Barclay, favoured this method. Some artistes to benefit from this technique included Shirley Bassey, Petula Clark, Denis Lotis and others, and it was particularly useful for multiple piano composite overdubs, for the likes of Joe "Mr Piano" Henderson and Winifred Atwell. Joe was a true recording balance engineer and an imaginative artist long before the importance of the sound engineer in creating the sound was recognised in the industry. It could make or break an artist, or be the difference between a hit or miss.

As mentioned earlier, the recording consoles at IBC in those early days had NO equalisation on the channels, from my hazy memory I think there were four portable equalisers that could be patched in – unimaginable to believe now. Joe made records that sold, and that was why he was in so much demand by producers, arrangers and artists. It truly upset Allen Stagg and some of the others who were steeped in what I now refer to as distant microphone placement, with much room acoustic heard: good for some classical music but a disaster for pop. I was privileged to be part of that learning. It stood me in good stead when I joined Joe and Denis Preston at Lansdowne Recording Studios a few years later, after my spell in National Service.

When on the road, Joe took no prisoners and if you had to play in music when recording a show and missed the cue or was late with the cue, he wouldn't let you forget it over dinner; but the next morning it was if nothing had happened. He was temperamental and explosive, and sometimes impossible to work with in the studio. If he shouted at me, which he did on a few occasions, I simply shouted back. I had much respect for him and I think that is why we got on so well and I learnt so much. There was criticism about the boss Allen Stagg, some of merit some not, however Allen recognised potential talent for the industry when he saw it and many of IBC engineers became much sought-after, highly talented engineers in their own right including Eric Tomlinson, Keith Grant, James Lock, Jack Clegg and Ray Prickett, Sean Davies (I know him and that is his right name) plus on the technical side Peter Harris and Ian Levene. My path with Peter Harris would converge much later in the late 80s, at CTS.

I recall one session for Johnny Franz, with Wally Stott arranging, although I can't remember the artist. It was a big band backing the singer to include four trumpets. I had set up the studio and Joe asked me to put two pencil mics (STC 4037s) one each between two trumpets – I thought, how can he do that, there are not enough input channels? He told me to go to the workshop and ask the guys to make up a parallel lead – two inputs to one output – it caused some raised eyebrows with the guys: "Can't do that – it'll be a mismatch for the input transformer," they countered. Well, regardless, and to their credit, the lead was duly made up. The session went ahead and the trumpet sound was phenomenal – the whole band sound was in your face but with perspective. It worked technically. Later in my career, I frequently used that technique if we were short on channel inputs especially for strings paralleling the mic inputs. It worked without degrading the sound; the gear was better able to cope with that in later years. IBC had various record companies as their clients using the facilities and we often ended up recording the same song again on an evening session with a different record company artiste. Joe was the "hit record man" and the companies knew that if Joe engineered the session it could well be a hit. Joe was conflicted and, I thought, uncomfortable with the circumstances. The situation finally reached a head when Joe was requested by two companies, Pye and Philips records, to record "The Garden of Eden." Pye recorded the song with Gary Miller, and then Frankie Vaughan sang it for Philips records, which he did. Joe didn't want his recording "secrets" (techniques) passed from one rival record company producer to another rival producer. Joe had his favourite producers with whom he liked to work; there were some he would not work with. Paranoid maybe, but simple logic in Joe's mind. Eric Tomlinson, a first-class experienced engineer, took over some of that work.

Joe would work incredibly hard, frequently working all night for his clients, which meant he came in later in the morning. I was in the studio entrance hall one morning and Joe had been working all night. Joe waltzed "or minced in" as he called it. Allen Stagg was there waiting for him to show up. When Joe arrived at 11.00 am,

Stagg pulled his left hand jacket sleeve up, looked at his watch and said sternly, "You're late!" – and then ranted on about time-keeping. Joe quite rightly ignored him and walked by to the lift. I think Stagg had a hang up about time keeping. It didn't seem to matter that Joe had worked most of the night. Joe was as upset by that approach as he was by the continuing innuendo from others behind his back because of his sexuality. With a client he was not a clock-watcher. He would put his all into the job regardless of how long it took – I doubt whether this was good financially for the company in fairness to Allen but we had to have the income. Did Joe charge it all on the timesheet? I don't know. It certainly was a problem later at Lansdowne (see chapter 5). Allen always encouraged innovation by others to maintain the very high standards set. Joe was the biggest innovator of them all at IBC. Yes, Joe was a loner and certainly not a team player and it got up people's noses but his results speak for themselves!

For mobile (remote) recordings, IBC had a BTR2 portable recorder so called but came in two green Rexene cases, tape transport and electronics linked together by large multi-pinned plugs. The machine's two boxes weighed 363 lb (165 kg)! No mean task to lug around! One occasion I remember Joe was taking the train to Scotland – no assistant – with this machine to record Jimmy Shand and his band. When Joe arrived the machine did not work! Joe effected repairs and came back with the recordings, somewhat of a rewire job in typical Joe fashion.

Towards the temporary end of my on-hold career to go into National Service, I used to go with Eric Tomlinson to record a show for Radio Luxembourg from the Embassy Club near Piccadilly Circus. It was a show recording with a small band and vocalist. On one occasion, I accidently plugged our monitor speaker into a mains feed on the equipment – there was one almighty bang and hey presto no speaker, cone blown out, fuses blown and a ringing left ear for me!

I was not flavour of the month with Eric, but fortunately we had a spare monitor. The problem was that the speaker cable had the same connector as the mains connector on the gear. I thought of that event and how dumb to make all the connectors similar. I never made such a mistake again!

Eventually I was asked by the club to manage the recordings using my own kit – the show usually started around 9.00pm. Allen Stagg got to hear about this and one time when I was about to leave the studio for the day to go to the Embassy Club to set up before the punters arrived for the evening; he gave me a large menial clearing-up task and offered to pay overtime, as my time ended at 6.00pm unless I was on sessions. I did the work in good time and reported back to Allen I had finished the job – nothing he could do, but he clearly wanted to prevent me from working in my own time for extra money. Nothing more was said after that.

On another occasion the company had a job to record local artistes/singers, pianists, harpists at an Eisteddfod in Aberdare, Wales. There was Eric Tomlinson and his wife Pat – a lovely lady – myself and Allen Stagg driving the old company van which rock 'n' rolled all over the road. We started out in the late afternoon from IBC expecting to be in Aberdare within a few hours. Then we had a water leak from a radiator hose… This meant stopping every now and then to get some water from wherever we could at a late evening hour, to top up the radiator – steam billowing everywhere from the faulty hose in the radiator. We spent the night in the van and in the morning a garage was finally found but they had no spare hoses for this old vehicle so we had to wait the best part of a day for a spare one to come from miles away. Van finally repaired we eventually completed our journey.

We had an interesting week in the Chapel Hall recording all sorts from Welsh harps to vocals and piano. Eric, Pat and I stayed with a miner and I remember how immaculately kept his house was and the warmth of the hospitality – such lovely, friendly people. I have never forgotten their hospitality. The husband took us down into the depths of the mine – a completely new experience for a London boy! My enjoyment of Aberdare was cemented by meeting a lovely, well-endowed Welsh girl with whom I enjoyed a quick snog one evening behind the church hall!

For those that are interested our recorders were portable EMI TR50s (two heads – erase and record/playback) again the case covered in that EMI green Rexene. After 59 years, I cannot recall what happened to the recordings or the lovely young lady from Aberdare!

Tape and Technical

Tape editing in the '50s at IBC was done with non-magnetic scissors, the reason being that if scissors that could be magnetised were used when the edit passed over the replay head there would be a "plop" – magnetism passed to the magnetic coated tape. The location of the editing in and out points was achieved by moving the tape by hand with two hands on the tape spools and moving the tape across the replay head to locate the beat or whatever part of the music had to be located – called scrubbing, and much easier to do when the recording was at 30ips – twice as much editing room. The in/out points were marked by chinagraph pencil, yellow or white, the tape removed from the head and held between the thumb and forefinger. A cut was then made at a judged 45-degree angle and the edited parts then spliced together with splicing tape. It could be a very labour intensive job particularly if the artist/s made mistakes as they often did – we had no editing blocks at IBC. With tape you had to get it right first time or you could be in trouble. Ironically the earliest digital recording systems, introduced in the late 1970s, actually made no provision for editing. The data was stored on videocassettes and you certainly couldn't cut the tape, so we had to wait a year or two longer before digital recordings could be edited. The first digital editors, made by Sony, used a control console which allowed you to roughly locate your edit points and you then used a 'wheel' to scrub a rather low quality version of the sound which had been stored in the console's primitive memory chips. Once you had identified your edit points you could then instruct the console to play you a high quality version of the edit in "rehearse" mode. Once all concerned were happy with the proposed edit you then committed it to tape and continued on to the next edit.

Digital editing immediately gave us one enormous advantage. When going from one take to another edits would sometimes be noticeable because the levels or intensities of the two takes were different, but with digital editing this could easily be corrected.

Once computers became powerful enough to handle digital audio the editor's life became easier still. We were freed from the restrictions previously imposed by tape and the computers could also provide us with a detailed visual representation of the sound. Once you could see the waveform it became possible to work with much greater precision. Another advantage was that edits once thought "ambitious" or "impossible" could often be made to work as each edit became a cross-fade which could be of almost any duration so that one take could be carefully blended into the next. Editing also became non-destructive; if anything didn't work you simply pressed the "Undo" button.

While I was at IBC, I studied electronics at London Polytechnic (evening courses) for an HNC in electronics. It was cut short by my call up for National Service and I never continued. I regretted it then, but not later on.

Mono to Stereo – Blumlein's Legacy

Discussions about microphone placement, and about innovative approaches to recording, such as Joe's, would do well to acknowledge the part played by other pioneers, particularly Alan Blumlein. His work set the scene for the arrival of stereo, and its eventual evolvement into surround sound. It's worth putting Blumlein's work into historical perspective.

Binaural Sound

Alan Dower Blumlein who is credited with being the inventor of Binaural Sound, was originally employed by Standard Telephones & Cables. He was a highly talented engineer and scientist working on several projects for the company improving various fundamentals of submarine cabling systems and designing various electrical circuits. The most important of which was the "AC Bridge Circuit" credited with earning ST&C much revenue in future years.

In February 1929, aged 25 Blumlein secured a job with the "Columbia Graphophone Company" in London. He was assigned by the company general manager, Isaac Schoenberg, to find a way around the Bell Telephone Laboratories Patent for a recorder/cutter system for cutting wax discs by designing and building a completely new system. Alan Blumlein had an immense interest in music, especially Beethoven.

May 1929 found Blumlein joining the team to work on the cutting system: before then he was working on other projects including a high frequency, high quality microphone project. At that time HMV Gramophone Company and Columbia Gramophone Company (two separate entities) were selling between them 30 million units per year and paying substantial sums of money to an American rival for the privilege of using a superior wax disc cutting system. It was necessary to pay a royalty to Bell Labs for every disc cut using the Bell Labs system of between 0.875 pence (21p) and 1.5 pence (32p) for every record cut. The recording equipment at the time was rudimentary and not up to the job bearing in mind that only a few years earlier recordings were made by an orchestra or singer performing into a horn driving a direct cutting system to wax. Schoenberg wanted to find a way round the Bell Labs patent and under Blumlein's electrical expertise aided by a team of engineers one was found. After much experimentation, and overcoming frequency response hurdles, measurements and comparison between the Blumlein system and the Bell Labs system showed "there did not seem to be any discernible difference between the two systems".

The Columbia recording system patent Nos. were 350,954 and 350,998.

"The Columbia Gramophone Company" (HMV) was in established in 1931 from a merger with the Columbia Graphophone Company. The new company was called Electric and Musical Industries, or EMI as it became known.

After this success, Blumlein moved on to develop his binaural sound system.

The Binaural Sound and Film Experiments 1931-1935 EMI Factory Blythe Road, Hayes.

Alan Blumlein applied for his famous Binaural Sound patent, No.394,325, on 14th December 1931; 22 pages long with 11 supporting diagrams and legends. He clearly laid out the principals of Binaural Sound, today we know it as "Stereophonic Sound" or "Stereo" for short. In simplistic terms two ears, two microphones, two loudspeakers reliant on the fact (well documented) on the phase, intensity and dependent on the frequency transmitted being differences arriving at the ears to identify the location of the sound.

Blumlein also described a method of reproducing sound for cinema – this was quite some years before Dr Ray Dolby's cinema systems – in which the sound would "follow" the actor as he/she moved across the screen. Blumlein and his team experimented with stereo-sound film cameras. They spent many weeks on location in and around Hayes, summer 1935, filming all types of subject material. The most famous footage is "Trains at Hayes Station", in which a steam locomotive was recorded as it left the station, the familiar chuff, chuff, of the train moving from right to left in the loudspeakers as the train moved away from the platform. There were further experiments with actors with the binaural sound recorded by sound film cameras. *Ben Hur* in 1959 was the first major motion picture to use multi stereo format system shot on expensive 65mm Eastman colour film wide screen format and released in 70mm six-track stereo. The extra 5mm of film between the 65mm negatives and 70mm prints was comprised of 2.5mm outside the perforations on either side of the film, allowing for stripes of magnetic oxide coating carrying up to six discrete channels of audio, an enormous step-change of advanced sound, extensively better in comparison with the mono optical tracks on 35mm prints of the time.

Unfortunately, the cinema speaker systems were not always up to the job of reproducing the high quality magnetic multi-track sound. Some were of the old horn type, which sounded "honky", and the frequency response was not always good either. It must not be forgotten these were the beginnings of improvements for the cinema sound experience; for Cinemascope, Cinerama and 70mm print releases.

Alan Blumlein was a modest man and always credited the work carried out by his colleagues. It was Blumlein's genius that invented the concept of binaural and, with his colleagues and their work at EMI made it possible. Many years passed before EMI would bring back to life Binaural (stereo) sound. Surprisingly, they discovered that much of the modern surround sound system used in cinema and home entertainment was laid down by Alan Blumlein in 1934!

Blumlein was far ahead of his time, and not fully appreciated until 20 years after his tragic early death in June 1942, aged 38 years, when the Halifax Bomber he was in crashed in Bicknor, Herefordshire.

Blumlein had been testing an experimental and most secret airborne ground centimetric magnetron scanning radar system (subsequently known as H2S). He was also credited with developing his own stereo-cutting equipment by means of a vertical-lateral technique using a stylus that vibrated in two directions; first recording one channel of sound in a groove laterally and then recording another channel of sound in the same groove vertically. After the initial experiments at EMI auditorium (Hayes) Blumlein moved the equipment to HMV studios where he and his assistants did two days of music recording which was deemed to be successful and decided to be more adventurous using this technique to record Sir Thomas Beecham conducting the London Philharmonic Orchestra during rehearsals of Mozart's *Jupiter Symphony* at the HMV studios, Abbey Road in January 1934.

In parallel and, I believe, unknown to Blumlein, experiments were being conducted by Arthur Keller and Irad Rafuse with stereo cutting in wax a single groove stereo recording at Bell's Telephone Laboratories in New York City on 1st June 1934: they felt it was "rather amateurish"! (Source: Keller and Rafuse).

In earlier experiments, Keller and Rafuse had already conceived a way of separating high and low frequencies and recording them on parallel tracks on the same record. Later they found a way of recording two entire sound tracks and reproducing both tracks simultaneously using a single pick-up. From this came two full-range bands to 10 kHz from left and right microphones in the same groove. Some say this was the birth of HiFi recording. The stereo technique was patented by Keller and Rafuse in 1938. (Source: Keller and Rafuse).

Many years later Dr Ray Dolby (Dolby Laboratories) introduced in 1975, a realistic 35 mm stereo optical release print format – some 40 years after Blumlein's experiments with stereo. In 1982 Dolby introduced Dolby Surround Sound, an advanced method of encoding stereo sound encompassing surround sound, as we know it today, for film with the sound optically encoded in between the sprocket holes and the picture frame. In 1988 Dolby Stereo SR was released (as opposed to Dolby Stereo with A-type noise reduction) for *Robocop* but this was a continuous analogue and printed between the picture and the sprocket holes, where the mono track used to be. Dolby Surround Sound is a generic consumer term but doesn't actually refer to a specific technology. Dolby Digital is digital 5.1 data printed in block between the sprocket holes. Dolby Digital was released on *Batman Returns* in 1992. Today further major advancements have been made with the advent of the digital projector, the sound and picture being delivered by a computer server, not only lateral surround sound but a vertical sound system known as Dolby Atmos® where the speakers in the auditorium are positioned over the listeners head as well. What an achievement from those early beginnings! I consider Alan Blumlein and Dr Ray Dolby both pioneers to achieve in audio a better listening experience for the consumer.

I was privileged to become acquainted with Ray Dolby from his early years at Battersea, when the company was a fledging in 1965. I frequently saw him at AES conventions in New York and Los Angeles, a charming man and Dagmar his wife, both with no airs or graces. He passed away in September 2013; the company lives on with his legacy.

I recall a conversation I had with Head of Sound Tony Lumpkin, at APBC (Associated British Picture Corporation) in Elstree, when I was invited by Laurie Johnson in the early 60s to visit one of the dubbing theatres and sit in on a morning's dub of one the *Avengers* episodes. I recorded the main titles and end of parts at Lansdowne. With Tony, the conversation got round to what was, in my opinion, the dreadful quality of optical sound in picture houses and dubbing theatres, and I thought it could be improved. I understood about the Academy Curve for optical sound (restricted frequency response) and how "locked in" the industry was at that time. Dubbing theatres used the academy curve in the monitoring chain for optical prints as standard practice.

However, I was sure the speaker systems and amplifier performance in picture houses could be improved dramatically, as well as the acoustic quality of those houses, as many were using old pre-war installations with no thought to sound quality. Tony's argument was that it would cost too much to change the standards in all theatres and the acoustics. Point taken. I suggested that surely we could get away from those old "peaky" horn-loaded systems with better speaker systems and amplifiers, and that there must be an improvement of the sound despite the academy curve roll off, "Why not progress as we are doing with monitoring systems? " Tony was well respected, and was a charming man of technical excellence, but I do not think he thought it feasible. I disagreed, and we left it at that.

The description of the Academy Curve, also known as the Normal Curve, is defined as flat response between 100 Hz – 1.6 kHz, with the response down 7 dB at 40 Hz, 10 dB at 5 kHz and 18 dB at 8 kHz! The purpose of restricted bandwidth was to reduce the noise floor from the optical track. In layman's terms, this is an extraordinary attenuation of the high end, and a serious reduction of the bass. Hardly HiFi! Today, we have the Tom Holman THX theatre speaker systems and acoustically well-designed picture houses, along with the remarkable advances by Dolby Labs that make the cinema big screen experience worthwhile.

My First Stereo!

At IBC we often worked for American clients, and on one such occasion the client brought with them a portable stereo recorder made by Ampex, it was an Ampex 600 machine housed in a sturdy brown case. The case broke in half, the top half contained the tape deck while the bottom half housed the recording amplifiers with two integral speakers. The band was the Ted Heath Band with Eric Tomlinson engineering. I was asked by Allen Stagg to record the band that was in Studio A in stereo, on the American machine, and myself on the studio floor. Talk about being pushed off the deep end! There were two inputs only going to left and right mike inputs on the machine. I thought the only thing I could do, given the circumstances (mikes were in short supply), was to use two spaced "Ball and Biscuit" microphones (the omnidirectional STC 4021 moving coil microphone, it had an accurate omnidirectional polar response for its time) spaced about three feet apart moving them around the studio until I got what I considered was an acceptable sound. I recorded "blind" so to speak, monitoring the band during recording on the VU meters making sure they didn't hit red too often and listening to playback after the recording via the two inbuilt speakers.

The results sounded good for just a two-mike pick-up. On playback Eric was happy, I was happy with the result and other engineers gathered around to listen, the Americans seemed pleased. I was told later it was an experiment for the Americans. I heard nothing more until I received a letter of thanks from Allen Stagg telling me the clients were very happy with the result. The band had a very good internal balance between the instruments – I got lucky! I wished I kept that letter, as I was obviously doing something right. I never found out what happened to that recording. It was the first stereo recording for me, and a first for IBC.

Chapter 4
National Service – Brylcreem, Boots & Blisters

Things went on progressing well with IBC, and I continued to learn as much as I could, until I had to leave the studio in 1957 to do my National Service. It was inevitable, and I had been fortunate enough to defer it twice with the help of the company. I resented the fact that my career was going to be forcibly interrupted for two years. A medical was mandatory before one would be accepted, and I went to mine in a government building in Acton – green walls painted halfway up, green canvas chairs to sit on, a strange smell about the place – then the medical check.

"Take off all your clothes, leave them there," A pile on the floor, in your birthday suit with many others, and tables with men in white coats sitting at them, doctors I presumed. Chest X-ray, listen to the chest, check eyes, ears, measure height, look at the whole body then, "Cough!". I went from table to table, examined by one white-coated person after another. That was it! It was a comic formal procedure, performed like everything the military did, strictly according to King's regulations. No result was given – just instructions to wait for a letter! I requested the RAF, Navy or Army in that order. The brown envelope with the confirming letter arrived sometime later. Army. As least I got my last choice!

I had to go to camp at Catterick in North Yorkshire to join the Royal Signals – Catterick Camp is still one of the Army's largest UK training facilities. I arrived on a Thursday – all new recruits appeared to arrive on a Thursday – no idea why! We lined up in our civvies then came the short-back-and-sides haircut, and *I mean* short! Those army barbers used old-fashioned large hand clippers, job done in less than five minutes, they loved their sadistic jobs those guys. To me, it reminded me of sheep shearing, with all of us waiting our turn and masses of hair strewn over the floor. We were proud of our hairdos in the fifties, using Brylcreem, oh yes sir! No individuality here though. We all had to look identical yet we somehow retained our personalities in barracks.

Then all the kit was doled out: tunic one, trousers two, boots two and so on through the whole kit in that army back-to-front jargon, "Sign 'ere!" And last the obligatory kit bag and army number given, "Remember it by 'eart lad." The uniform was of some lousy material that made you itch like hell! I was assigned to Four Troop Royal Signals. We did the ten weeks square-bashing marching that you did in the 1950s, with the constant bullshit of the kit. The only thing it gave you were awful blisters on your feet. I learnt a new language. "Blanco", "spit 'n' polish", "rifle oil", "pull throughs", "jankers", s*** house and the dreaded "bull", especially the boots that took ages to get a mirror shine – they were issued in raw black leather. The basic training was carried out by corporals. Most unpleasant malevolent figures who treated us like dross! Not to mention the endless inspections parades, and time on the firing ranges using a STEN gun with one loaded clip, and the .303 rifle with live ammunition, ("Sign 'ere five rounds only"). We were driven out to the bleak North Yorkshire moors in all sorts of weather.

I thought the whole thing was a farce, especially polishing the barrack floors with a heavy hand-held oblong polisher on a broom handle. Where were the polishing machines? "We don't have those, plenty of sprogs to do the job"! Typical Army! We thought they'd have us paint the blades of grass white with a toothbrush next. Getting out of bed very early was decidedly unwelcome – and, oh, I soon got fed up with it! Grit your teeth and smile was your only way through it!

Every day the next day's orders were posted on the notice board – a must-read affair, written in typical Army jargon. The food was dreadful, doled out by army chefs – were they ever trained? I doubt it! We moved down the queue and a portion of food was plonked on, the next position another portion and so on – ghastly. Tea time was just as bad, always bread and other unappetising things, dreadful jam out of huge tins. We all sat in a large mess hall and ate out of our small mess tins, which were used for main course pudding and anything else you ate – no plates – the washing of which consisted of a large tank of hot water, which you dipped the tin in and let it dry. The scum and grease round the edge of that tank was something to behold, as hygiene was not considered. Today, it would be condemned. I more than frequently had an upset stomach. An officer would come strutting around from time to time walking down past the tables, "Any complaints?" No one complained because if they had, there would be trouble. The NAFFI did well in the evenings.

Every month or so we had to do the obligatory guard duty walking around the camp with pick-axe handles. This included checking the rifle store – all three floors with a stench of gun oil and watching over the railway line into the camp. During the winter, the four of us on duty would spend the night in a platelayers' hut, which had a wonderful coal fire, and we could get some sleep without an officer or a Senior NCO poking his nose in – it was too far for them to walk! They were lazy gits! The first troop train arrival was about 05:30 over the weekends carrying troops back from leave. In my mind, another useless and total waste of time – only in the army in those days! If the camp was going to be hit, pick-axe handles would be no good!

I was a junior NCO Lance Corporal and if anything needed to be done, the order was passed down the line to an NCO who passed it on to a junior NCO. Most of the Senior NCOs and Warrant Officers were up where "the-sun-don't-shine".

On more than one occasion, there was an Orders of the Day notice asking for volunteers to go to Porton Down, with the promise of extra pay and better conditions than camp plus it meant a week off. It was said they were looking into a cure for the common cold – I thought oh yeah! I knew of Porton Down, it was where they conducted various chemical warfare experiments alongside other secret work. Asking for volunteers occurred more than once. Many of the guys went from my troop and came back with tales of drops up their noses and on their skin. Others were put into chambers filled with gas substances. Some came back feeling not right.

Many years later there was an advert in one of the papers asking for people who had been stationed in Catterick in the '50s to come forward. It transpired there was an action against the MOD as some of the guys in later life became ill and died, and the affect on their health had been traced back to Porton Down. The government wanted proof of these posted Order of the Day notices. I came forward and explained to some Whitehall office person that I had been in Catterick and these notices were a regular occurrence and some of my friends went to Porton Down. "Do you have a copy of the original notice?" was the question I was asked.

That was a bloody silly question to ask after nearly 50 years, anyway it would have been more than my life's worth to remove one. "We need the proof," was the answer. Oh what a surprise, the Army never kept copies of Orders of the Day that far back. Another sensitive secret military whitewash? I heard no more. I guess the standard answer was "not sufficient evidence" to prosecute those responsible for the horrendous experiments.

We now know that some 20,000 servicemen were duped into volunteering for research into the common cold and then used in the most atrocious experiments with nerve gas and other things. All I can say is thank God I didn't volunteer.

The square bashing trainers were Sgt Toole who was full of himself, Cpl Harbron who took great delight in shouting all the time, much to my amusement, although the smirks I kept to myself while parading around on the parade ground with an old WWII .303 rifle that had seen better days. The NCOs especially the Sergeant Major seemed to be manic psychopaths, who positively enjoyed shouting at, and insulting, us new recruits. L/Cpl Thomas and L/Cpl Meadows were pleasant enough chaps but in my mind nonentities. I guessed they didn't like National Service any more than I did.

After the square bashing there was the passing out parade, which my father attended. I was delighted that he came all that way – my mother was busy at home – we had a good time together and I was pleased to see him because during all the weeks of square bashing no leave passes were allowed unless on compassionate grounds. It was my first time away from home, which I missed greatly.

After being allowed forty-eight hour leave passes and, some days later, we were all lined up to be told where we were going next. I was a bit anti all the fighting stuff and my dad certainly didn't want me to go to Kenya, where many were sent to fight in the Mau Mau Uprising, which had begun in 1952. When I look back, so many lives were lost for what!!? Afghanistan and Iraq today comes to mind – all those lives lost and many wounded veterans!

Many of the guys volunteered to go to Kenya. I think the army had enough guys, so none of us were pressed ganged into going, but persuaded by the travel and excitement of going to a foreign country at the army's expense.

Some of the guys were automatically selected to join the Cypher section – how the selection worked was a mystery, and were sent to a Cypher training school in Brighton. "Lucky buggers!", we thought, to be beside the sea. A cushy job for those bright sparks from Britain's public schools. The class system ruled so not a hope in hell for something so easy for a grammar school boy.

I happened to learn on the "gripe vine" that the camp was looking for teleprinter trainers. At that parade of guys, the question was asked who would like to stay at Catterick to train teleprinter operators – silence, no takers except me. As they say, I got the gig. It turned out to be pretty cushy. I was automatically made up to L/Cpl, going up in the world, I thought, "Two stripes next…"

"Dream on sonny!" as my superiors would say.

As a L/Cpl, it was one of my duties to march the chaps in training from the training huts to and from the barracks, in the morning, at lunch time and again in the evening. About one mile, in all weathers, and when it was raining, we had to wear dreadful rubber army capes that stank to high heaven especially when wet. Humpty Dumpty came to mind except there was no hill to march them up or march them down.

The teleprinter operator's course was quite intensive and lasted for some twelve weeks. We had to learn by touch-typing with a cover over the keyboard and were tested at every stage of the course. You didn't fail! The final phase of the course was to learn about how teleprinters operated mechanically and electronically. They were electromechanical devices, sending on and off pulses at 75 words-per-minute from a punched tape or directly as one typed not at 75 words-per-minute! The technical side was right up my street. In the training huts was a Corporal Bulgin, a decent chap with whom I became acquainted.

I became responsible for the last four-week phase as a trainer for the tech stuff, teaching these guys the ins and outs of the teleprinter, and the signals it produced. I also ended up in charge of the camp cinema, which suited me very well! Ah, but before I was allowed to touch the old clapped out 16mm projector that had seen its last days, I had to attend a course at another camp – can't remember where. It was a good week's skive – just to learn how to project films! Only in the army! I should mention here that another part of our training was to learn to operate a microwave communications speech transmission system, a cumbersome device for use in the field. It took ages to set up and then only worked if pointing on a line of sight, as microwaves don't go around corners. Well, at least they didn't then! Can you imagine setting up a system like that in the heat of battle? It was deemed secure as no enemy could listen in unless he had a receiver lined up to your transmitter. Hmm... I thought the whole purpose of Cypher was to send encoded radio signals that the enemy was thought not to be able to crack.

However, all was not doom and gloom at Catterick, as there was a lovely Yorkshire town nearby – Richmond – a pleasure to visit and spend time there. I joined an operatic company, and kept up my singing with the group; with rehearsals every week it was good fun. We did *The Pirates of Penzance* for one of the shows. There was a girl in the cast about my age whom I thought had a penchant for me. On one occasion I was invited to her home for dinner, some way out by bus from Richmond into the beautiful Yorkshire countryside, a wonderful dinner with her, me and the parents and a coal fire in the grate. What a change from camp food and cold barracks! After dinner, her parents left the room and we sat by the fire chatting. It was a splendid evening. I learnt that her father was the local chemist of some standing and I was taken back to Richmond in her father's car – quite a treat then! When I left Catterick and the players, we lost touch.

Weekend 48-hour passes were good to have; from Richmond Station direct to Kings Cross return on the troop train was 32 shillings and 6 pence (today £38.09p). Being a troop train, it was pretty slow so in the end I gave up and travelled to Darlington Station for the London Express to Kings Cross. On a Friday I could skive off early, no one noticed as they were all at it, officers too!

Going back to camp on a Sunday evening was a slow and cumbersome journey on the 23:30 troop train from Kings Cross directly back to camp. In the winter the steam heating of the carriages worked sporadically and the train stopped frequently for no apparent reason. Peterborough seemed to be the favourite – arriving at camp at 06:30. If I was lucky and the corridor train was not too full, two of us would share a whole compartment, pull down the carriage blinds and stretch out across the seats for a good sleep, hoping no one else had the same idea.

I spent the day of arrival skiving or doing practically nothing, especially if I had not been able to sleep on the journey.

I also took advantage of external courses that the Army had on offer for us – Signals, for example. I attended a long course at Newcastle University learning about microwaves, wave guides and microwave transmissions. They had a computer called Ferdinand, (a Ferranti Pegasus – **FER**ranti **DI**gital and **N**umerical **A**nalyser **N**ewcastle and **D**urham. (Source: school of computing science Newcastle University) which I was hoping to see, but to no avail.

Another part of the job was to be a sort of secretary; I worked in an office, an old wooden army structure similar to the training huts. I had a sergeant in charge of me at the time – a chubby, very funny guy, for whom everything was "tee'd up, squared off and tickerty boo" – think he must have been a draughtsman in an earlier life! He was a career sergeant whose turnout was immaculate; you could see your face in the toecaps of his boots as in a mirror. I took instructions from the sergeant, who in turn took orders from the officer in the hut located in a separate room – *Dad's Army* comes to mind now – but it was true. Every day I had to type on a stencil (can't make mistakes) the next day's orders and then use a Roneo stencil machine to hand crank the orders off onto blank paper. A very messy job, as the black ink went everywhere. I used what I now call an old fashioned clunky typewriter, like the ones seen in old movies. To be able to do all this work, I was made from Lance Corporal then acting Corporal, so I dreamed on sonny! I never received the other stripe as I was shortly off to London courtesy of Desmond Beatt and the MOD. We used to get a break to go to an old NAAFI van (must have seen war duty) every morning, which I looked forward to, for a morning roll and a coffee. The two women at the counter were very affable Yorkshire lasses, greeting you every morning with a cheery "What'll it be this morning luv?" While I was at Catterick, I kept in touch with an ex-army Major, Desmond Beatt. I believe he was stationed in India during the war. At IBC I had been in attendance on many of his Carlton Facilities sessions with Joan Walker producing. I contacted Desmond; he knew quite a few people in Army offices in Whitehall, and pulled some strings to get me transferred from Catterick Camp to BFBS, the British Forces Broadcasting Service based in Smith Square, London for the last six months of service in 1958. Ah, that was more like it!

The nasty piece of goods Corporal I met years later, in the mid '60s, when walking up to Notting Hill Gate for lunch. He was with other people, outfitting a shop, and looked a very different person when I went over to him. I recall saying "Hello – do you remember me?" A blank stare and then he remembered. I said I remember you very well and that was it. He knew.

Corporal Bulgin's family, it turned out, was behind the Bulgin electronics company still in business today.

At BFBS there was one studio and a copying room, managed by an Irishman named Paddy! He was a nice guy and just let me get on with it, I rarely saw him. The studio had small mixer with very few inputs – can't remember the name – it worked, just!

For the first few weeks in London, I was stationed in the Guards barracks in Whitehall. Every morning at 06:00 hrs, the pompous Warrant Officer came banging on our cubicle doors with his stick and literally shouting at us to get out of bed. We had to jump to it and go running around Regents Park in all weathers. It was not to my taste! Get back, shower, breakfast and ready for the workday at 09:00 hrs. I walked to Smith Square, and it was a relief to work with civilians in a relaxed atmosphere. I was subsequently allowed to live at home where my father and mother had a sub-post office and general stationary shop at 54 Fulham Palace road – we lived over the shop. I commuted to Smith Square by Tube. Thursdays was payday at the barracks: you had to wear full kit for this ridiculous piece of theatre and be immaculate with all brass buttons polished etc. I kept my kit all Blanco'd up, hanging in my locker and ready to wear, to save cleaning it every week. The process of applying Blanco – to keep parts of your kit, such as webbing, a consistent colour – was a messy labour-intensive task, and deservedly unpopular! It really was a farce, but the regular Guards guys took it all so seriously. I only thought it happened in the films – evidently not!

We did a lot of recording at BFBS so I was in my element. Part of my job was to record radio dramas, and be creative with the sound effects – an enjoyable and rewarding process. The material we recorded was for distribution on tape to all our overseas forces bases – Army, Navy and Air Force. One of my jobs there was to copy this material at 7.5ips and put all the copies in a tin can to be sent abroad. When the tapes returned, I was expected to erase them and re-use the tape. That was fine in theory, but many tapes had been out in very hot and humid countries, and when you opened the can, the tape virtually fell apart – the backing had all but disappeared into a pile of loose oxide!

One day, I was told I had a recording to do – a small band. In those days, I didn't even know what an electric bass guitar was (shame on me): acoustic (upright) bass, yes, electric bass, no. With the band was a singer, Matt Munro. This guy comes in, a good singer, and we record tracks for the Forces' broadcast. After the session, Matt said would you like to come to the pub – our local, just off Smith Square. I hardly drank in those days, when Matt asked me what I'd like, I said a bottle of Toby Ale, a half pint that cost about one shilling and sixpence (7.5p), I think! Overall, it was a very pleasant recording session, and Matt was a really nice guy – didn't see him again though! But I knew of his later recordings at Decca, then EMI Abbey Road.

During my six months at Smith Square, Denis Preston came in to record a record production presentation, I can't remember what. He recognised me from the IBC days and asked what I was going to do when I was demobbed from National Service – I told him, "I guess I'll go back to IBC". Denis replied, "I shall be having my own studios now in Holland Park with Joe Meek, please contact me when you leave the army, I can give you a job".

I guess my musical background helped and I agreed with Joe's recording techniques – a very "present" upfront sound that made big hits for the record companies, and thus earning them and their artist's considerable sums of money. It was a change from the old "dyed in the wool" recordings of the past.

I learnt a lot during my spell at BFBS, gradually increasing my audio experience and trying to use what I had learned at IBC, until eventually I was demobbed towards the end of 1958. I had to go to Chester for that, a long journey by steam train, and stayed in the barracks for a couple of days. We were given our demob papers, and donned our civvy suits, which we hadn't worn for a long period of time. We were also given a rail warrant for the journey back to our home towns. Hallelujah – it was great to get back home out of the Army and that itchy uniform!

Lansdowne House – Historical Background (suggest using this in side-bar, with pic of building's exterior)

Lansdowne House, located in Lansdowne Road, in the Holland Park area of London, is a short stone's throw from Holland Park Tube Station. It is a large, eight-story block, with apartments on the ground and upper floors. It was designed by William Flockhart and built for South African Mining magnate Sir Edmund Davis in 1904, to accommodate various artists, including Charles Shannon RA and Charles Ricketts RA, who both lived and worked in Lansdowne House until 1923. Other artists living there included Glyn Philpot, Vivian Forbes, James Pryde and F. Cayley Robinson, with their homes in the various studios around the house. The original artists' apartments and work studios faced north, with high ceilings. In the early 1950s, the original apartments and studios were divided up and converted into flats, with the exception of Lansdowne Studios Flat One and one top floor flat. A lift of the open-caged type served the upper floors – elegant but unreliable! The winding stairway's handrail was continuous, solid polished mahogany. In the early years, when the artists occupied the house, the basement and sub-basement comprised a sub-basement squash court, a basement-level smoking room and a slipper bath, complete with coal cellars and a boiler room, plus a caretaker's flat and the utility supplies and meters, together with a garage in the adjoining mews. Lansdowne House was listed Grade II in 1969.

Chapter 5
Denis Preston, and the Launch of Lansdowne

Denis Preston was a valuable client to IBC as he brought a huge amount of work to the company, mostly jazz and traditional jazz bands. Joe Meek was his preferred engineer, because of Joe's obvious flair and the successful recordings he made at IBC. However, the problem was that Allen Stagg – probably due to his own resentment – had a preference for rotating engineers, even if they did not suit the client. In other words, Bookings made a decision without reference to the engineer or client unless the client made a specific request mostly for Meek, then it didn't always happen for Joe.

At IBC, as mentioned previously, when I was assigned as assistant on a session I asked Bookings who was the engineer. I was told the name, but knew Joe would not like a client who had requested him to be given another engineer. This infuriated Joe – and particularly Denis Preston – who had words with Allen Stagg and, as a result, Denis fell out with Stagg because of his bloody-minded attitude. Denis was spending in the order of £6,000 plus per year (£134,000 today). Joe's recordings for Denis at IBC were impressive.

A small sample would include, at IBC and then at Lansdowne with Joe and myself: Chris Barber, Lonnie Donegan *Rock Island Line*, Humphrey Lyttelton, *Bad Penny Blues*, (see Chapter 3, a massive hit for Humph), Big Bill Broonzy, Ken Colyer – a difficult person to get on with and steeped in the what we called old fashioned recording sounds. He was not interested in "modern" recording techniques so not up Joe's street; Monty Sunshine with *Petite Fleur,* Laurie Johnson, Kenny Baker's Baker's Dozen, George Melly, Mick Mulligan, Johnny Duncan with *Last Train to San Fernando,* Chris Barber's jazz band with Ottilie Patterson, Terry Lightfoot, Sandy Brown, Al Fairweather, Alex Welsh, Monty Sunshine, George Melly, Stanley Holloway, Stan Tracy, Elaine Delmar, Tony Coe, Roger Whittaker, Wout Steenhuis, with whom I spent many hours recording *Hawaiian Music,* Kenny Wheeler, Don Rendell, Ian Carr, Neil Ardley, Joe Harriott, Jack Elliot, John Dankworth, Cleo Lane, Dill Jones, Don Rendell, Annie Ross, Galt McDermott, Lansdowne String Quartet, Amancio D'Silva and Ghanaian drummer Kofi Ghanaba, Preston was a leader in the production of Calypso and West African music. The list goes on and on. See - https://www. discogs.com/artist/304822-Denis-Preston?page=1

Preston was a very prolific producer and his own man, some say arrogant. I never ever saw arrogance, although he didn't suffer fools gladly. If he had an idea for a recording he would talk about it with those he wanted involved, artists and balance engineers and depended on their artistic and technical skills for a positive outcome. We often met over a drink and sandwich in his office. I recall one occasion when RCA America wanted to release Roger Whittaker in the US, they sent a forty-four page contract! Setting out how Roger should behave in a typical Hollywood style contract. Denis was furious, "It's effing rubbish, who do they think they are, up their pipe!!?" he said, taking another gulp from his whisky shot glass and puffing on another Gauloise fag. Having been sent back, the outcome was a contract without all the crap and reduced to four pages! That was the Preston approach. The first recording date of many I had with Roger was *Mexican Whistler* in 1968. Followed by his hit *Leavin' (Durham Town)* in 1969, which was in the charts in December of that year. His vocal sound was rich and present and a joy to record usually using a Neumann U47. Roger wouldn't record without the U47. Many more were to follow in the 70s and that's for another time.

Before he founded Recorded Supervision (RSL), Preston produced for Decca Records as well as London Records, recording George Shearing and Josh White in July 1950 for Decca on the London Records label.

His other recordings at that time were with the Calypsonians *Lord Kitchener* and *Lord Beginner* in 1950, for the Melodisc and Parlophone labels.

In my opinion, Denis was a very foresighted producer; he founded his own record company, Record Supervision Ltd, in Newman Street, London on 14th December 1954, with the added objective of also providing a route to publish the songs or works in a separate publishing company. He had a deal with Pye Nixa "Jazz Today" series and when that contract expired in 1959, Denis entered into an agreement with EMI and the "Lansdowne Jazz Series" was born and released on the EMI Columbia Label. Norrie Paramour was EMI director for Columbia records label. Denis and Norrie were great mates.

Denis decided to establish his own studios after falling out with Stagg over the assignment of engineers to his sessions – he always required Joe. Denis saw Joe the engineer as a creative partner – he did not treat engineers as technicians in white coats. Equally, Joe pioneered the studio as an extension to the musicians and their instruments.

Denis Preston's patience with IBC ran out in early 1958. Without a guarantee of always getting Joe for his sessions, he asked Joe to find premises to set up his own studio to rival IBC, with Joe as his engineer. Joe found a studio in the basement of Lansdowne House, Lansdowne Road in London's Holland Park, just a stone's throw from Holland Park tube station. It had previously been used by an amateur cellist for his own classical recordings. More or less what we called then an "egg box studio" – old egg boxes used for acoustic treatment. It was a rundown mess. Many years before it was squash court. In the meantime, Joe left IBC, persuaded by Denis with the lure of his own studio, and went to Preston's office in Newman Street where Denis had his production company. The service agreement for Joe was prepared on 7th July 1958 by the company solicitors.

During my last six months of National Service when I was posted to London, I kept in touch with Joe. I always got on well with Joe and Denis at IBC, hence I was delighted when I met Denis at BFBS and he offered me a job, asking me to get in touch when was demobbed from National Service, which I duly did. I guess my musical background helped and I agreed with Joe's recording techniques – a very "present" upfront sound that made big hits for the record companies, and thus earning them and their artist's considerable sums of money. It was a change from the old "dyed in the wool" recordings of the past.

When in National Service, the government of the day made it compulsory for the company the serviceman had left to reemploy them after the two-year period of service. When Denis offered me the opportunity to join him and Joe at the new Lansdowne Studios I jumped at the chance! I told Allen Stagg I would not be returning to IBC – he was magnanimous in his response, and wished me well for the future. As a new studio Lansdowne was a fantastic opportunity – and it was more money!

Lansdowne Recording Studios Ltd was incorporated on 6th June 1958, certificate number 605871. Also in June 1958, a lease was signed for the new studio and the previous tenant bought out for a small consideration. Funds for the new studio were procured from Denis's acquaintances in the music business; Kenny Baker was one of the shareholders as he had some success with the Baker's Dozen produced by Denis at IBC. There were twelve shareholders in all. The Financial Director and shareholder was Lionel Stevens, and one other shareholder director was Lyn Dutton – I saw very little of Lyn Dutton. He appeared to be a sleeping partner and had other business interests, including managing The Spinners, and owning a picture gallery in Marlow, Buckinghamshire.

Sandy Brown, the chief acoustics architect at the BBC, master-minded the acoustics for Lansdowne Studios in 1958, 11 years before he founded Sandy Brown Associates with David Binns. Sandy was a multi-talented man; a prolific writer regularly writing jazz articles for *The Listener*. He was also a lecturer on acoustics, jazz clarinet player and a recording artist with his own band, often recording with colleague Al Fairweather for Denis Preston. As a matter of interest, together with Al, Sandy recorded *McJazz* with Preston, and the project was named in 1959 by *Melody Maker* as one of the twelve greatest recordings of all time. One of Joe's again!

The studio decor and office design was by Ian Bradbery (he also wrote sleeve notes for Denis Preston's product), with furnishing material by Terence Conran (now Sir), of Habitat. The building works were carried out by a small firm whose name escapes me, but they did a good job! The electronics side encompassed the all-valve (tube) recording console. TR90 tape machines – two mono, one stereo – which were manufactured, supplied and installed by EMI in Hayes. Lansdowne effectively became the first turnkey studio.

Joe specified the console with Painton (the manufacturer) quadrant stud-type faders, and it had good basic equalisation with a very large Peak Programme Meter in the centre, 12 channel inputs and two-channel output – no panning. The colour was maroon with gold edging – typical Joe flair again!

While Lansdowne was being constructed, Joe set about editing the work he had recorded for Denis at IBC in the Newman Street offices. Before the EMI custom console was installed and while the main studio was being constructed, Joe used an old console from Newman Street (Preston's Record Supervision Company) for his editing and master copying work, temporarily installed in a half-constructed control room. IBC Management could hardly bar Joe as Denis's engineer if it wanted to hang on to the work while it lasted – after all, Denis spent prodigious amounts of money at the studios and on musicians. Richard Preston on Joe: *"Joe Meek was the first producer in UK to establish the idea of the 'producer's sound'. He used the studio as an instrument not a machine in a manufacturing process."* True words from Richard. To me, Joe was a kindred spirit.

Denis and composer Laurie Johnson were good friends and in partnership together. Laurie first met Denis at IBC in the 50s. Laurie Johnson recalls: *"We met on a recording for MGM records in the 1950s with Bert Ambrose, with Denis producing it for them and I did the arrangements. I think I was under contract to EMI at the time. That's when I met him at IBC at Portland Place and we recorded there and that is when he started talking about us recording together, and he was starting up at that time as the first independent record producer in the country. We made several albums and became great friends. He was a great enthusiast and we spent a lot of time together with interests outside recording – we had a similar sense of humour. We recorded three or four albums at IBC and Joe Meek was there, he was the regular engineer."*

I recall that outside of their musical collaboration, Preston and Johnson were great Billy Bunter fans! Their sense of humour was very dry and sometimes cutting.

Denis said he was going to write Laurie's autobiography: Laurie again, *"Denis was going to call it* 'Give me back my contract', *because of the artists, and what they would say when they came in to audition. His favourite story was that certain artists, and I won't say any names, but they would say (affects cough) "I have got a bit of a cold today and this pianist is not my regular one and I don't know the song very well but it will give you an idea." Anyway the pay offs, making excuses, that was the one…"*

Having recorded Shirley Bassey's first demo, I mentioned this to Laurie: His reply: *"Did you know that Denis auditioned and turned down Shirley Bassey, as he said she didn't know any blues. That was his yardstick, whether she knew any blues!"*

So on Monday 5th January 1959, I joined Lansdowne much to the relief of my parents, who worried about what I would do for the future.

When I arrived, I wondered with anticipation, what I was going to see; here was this building site located in all the basement of Lansdowne House, coming together with the equipment already installed in a small control room off the studio floor. Adjacent to the control room was a small room, and the entrance to the studio floor, with a heavy interlined curtain forming another part of the room area, to act as a separation booth for vocals or other solo instruments. There were two corner Tannoy monitor cabinets in each corner, fitted with 10-inch Tannoy dual-concentric speakers, positioned either side of the small control room window – the speakers were driven by Quad valve amplifiers. I later changed these to the large Lockwood monitor cabinets, with 15″ Tannoy Reds.

A rack of EMI TR90 tape recorders, one stereo and two mono, were against the right hand wall, together with twelve microphone amplifiers in an adjoining linked rack. Their gain switches rose in 10dB steps to 60dB of gain, with a vertical wire ended valve level indicator with the calibration engraving on the front panel to show any gain overload. The amplifiers were finished in maroon to match the console, with each microphone amplifier easily removable for servicing. These mic amps ran very hot because of the valves being in a small confined space meaning no forced cooling. There was also a small ¼″ jack-plug rack for patching in various pieces of external equipment. The whole installation was very well thought out, in no small part thanks to the expertise of EMI, who had their own studios in Abbey Road – then called HMV Studios, and which featured Hayes factory-built equipment. The studio floor, with its acoustic treatment, was almost completed. There was a much larger room upstairs, with a view overlooking the studio floor, and I was told this was going to be a stereo control room, which was Joe's thinking in those days. In some ways I thought that was rather odd when the console downstairs had two outputs, but there were no pan pots; however, as a newbie it was not my place to question. Looking back now by today's standards it seemed primitive but then, it was considered pioneering days in audio.

Joe had ordered from a company called FWO Bauch (who represented Neumann in the UK) an amazing collection of Neumann capacitor valve (tube) microphones. These included U47s, a great sounding mic (utilising the now defunct VF14M tube), as well as U49s (AC701K tube), KM54s (AC701k tube), and U67s (EF86 tube, which I later changed for an EF806 – a gold-plated pin, sturdy military version, which had less microphony). Amongst the order was also an SM2 (stereo microphone – with 2x AC701K tubes) and, from STC, a collection of dynamic mics that we had at IBC, plus a RCA BX 44 ribbon. For echo (reverberation), there was a live room in the basement, with screened cables to the control room for sending to a loudspeaker and receiving the mono echo return to the console.

Denis always had pre-session planning meetings with his engineers, describing what recorded sound he had in mind, but he didn't tell them how to record it.

With large music productions, I always met with the composer to read through the scores for a better interpretation of what the composer or arranger had in his mind. It gave me a good idea of what was to come on the date. Something I never forgot was Denis saying, "Do your homework behind the desk then you know what to expect on the date". I wanted to ensure I knew what to expect from the scores, and sometimes I would ask for copy scores in advance, so I could go through them and make notes. It helped me to decide on the orchestral studio set-up.

Having done the preparation and got it in my head Denis left it up to me or the other engineers on session dates to get on with it. Denis was the producer who recognised talent and put together the right team, artist, engineer, arranger, musicians and studio. His productions were released by these companies under contract with them – usually three years then the masters reverted, a very astute decision. I believe Denis *was* the first person in the UK to credit the engineer and musicians, and he certainly was the first independent producer. He was a very prolific producer. Before becoming a producer with his new company, Denis had worked in the forties as a jazz critic and presenter on the BBC.

Denis told me, *"We don't manufacture hits... we make records, and if we're lucky, the record-buyers will make one of them a hit now and then."* How true that was with many of the artistes that Preston produced.

Richard Preston on his father: *"Denis never claimed to understand the technology. I think that he saw himself like a film director who made sure he had the right cameraman in his crew, someone who could capture what he had in his head, e.g. the 'fizz' on the snare on Bad Penny Blues."*

Denis was a far-sighted and astute producer who took no prisoners. He later became famous at Lansdowne for his mega outbursts – his Mark Is, Mark IIs and Mark IIIs, as he called them! The target for these outbursts were generally the musicians, some of whom were not terribly cooperative and worked by the unionised "book" that governed what they did on sessions. Denis detested the Musicians' Union (MU) and, as he called them, "their soppy rules". He particularly disliked Hardie Ratcliffe, MU General Secretary in the early 60s, and the union intransigence about session rules and overdub rates. We worked around it with musicians who turned a blind eye – the session cash in hand ruled!

We had affectionate nicknames for various musicians some knew those names others didn't: upright bass player Arthur Watts – "The

Caterer", guitarists – "Chinese Charlie", "Hairy Hands", Eric "I shall survive" Ford, Mitch Dalton – Doctor, Arranger/Conductor Johnny Arthey – "The U-boat Commander"and Banjo Billy. There was also a percussionist "Dapper Dan", and three drummers: "Big Nose", "Crem" and "The Olympic Drummer", so called because he appeared to get to the end of an arrangement before anyone else in the band. The string players, violin or fiddle players were collectively known as gypos, a cutting engineer at Pye studios called "The Mad Turk", a pianist/accompanist called "The burglar", a sax and flute player known as "the policeman", another saxophone player "spanswick" – don't ask! When in a session planning meeting, Denis would think about a rhythm section, and ask you, "What do you think?" He would say to his secretary Silvia Pentlow, "Get Ginger (Baker) and Jack (Bruce) on the phone!" She did, and they came; you didn't refuse DP, as other musicians didn't.

Denis knew Ginger and Jack from their involvement in Alexis Korner's "Blues Incorporated" band and when Alexis formerly played in other bands in those early days. He played with Chris Barber and Ken Colyer.

Engineer Vic Keary was the engineer who recorded the band at Lansdowne (returning as a client) two dates the second one in May 1963 with Vic producing.

Some musicians would deliberately play for overtime by querying notes on the parts about two minutes to go before the session end, thereby incurring expensive overtime even though they had already played the piece more than twice and got it in the can. However, Denis deemed requesting another performance (to have a choice) with two minutes to go achievable. There was one excellent trombone player who chanced his arm for the fifteen minutes of overtime, and so Denis used his Mark III outburst and told him he would never be employed again on his sessions! And he wasn't. That guy was a very well known superb player with a long session career and a fixer, who at one time played with the Glen Miller band when they were in England during WWII.

I recall a Roger Whittaker session, on which there was a sax player who started to pack up his instruments two minutes before the session end, causing the whole of the last take to be abandoned. His excuse was the union stuck to time and that included packing up! The other guys on the date looked on embarrassed, as Denis had a Mark III with the guy – the session was terminated with the loss of one more take. That sax player was never engaged again. All studios in London had to be on the union's "fair list" and if not, the musicians' fixers (contractors), of whom a handful had the London session scene stitched up, would occasionally advise clients who requested to book musicians and a studio, not to book a particular studio if the studio had fallen out of favour with the fixer. If a musician fell out of favour for whatever reason with a fixer you didn't work for months or again for that fixer. The really powerful fixers were usually violin players and led the orchestra if they were playing on the session. Denis saved money by not having to pay a fixing fee. He also used one fixer, David Katz, who happened to be a shareholder in Lansdowne! David always booked for Laurie Johnson and worked with Denis, when asked why a particular requested musician was not booked for the session replied, "He was not available". It later turned out when the musician was told, "Sorry you were not on our session". The guy said he was never booked and was sitting at home and available! David Katz (a violinist) never worked for Denis and Laurie again – he was sacked instantly. Denis was furious with David, who was a shareholder in Lansdowne, for "pulling the wool"!

He must have lost a lot of money because Laurie's orchestras were large, thereby attracting a fat fixing fee. Before David was dismissed, he used to score read in the box on sessions for cuing solos or other changes in the orchestration, because although Denis and I could score read, it took the pressure off Denis to produce and me to push the faders. It also allowed him time to have the occasional glug of whisky or brandy from his shot glass – he was an enthusiastic drinker – along with his Gauloise fag. A typical session with David would involve him saying, "I want to hear 10% more of this" or, "5% less of this section." And so he went on and on every take – it became a bore! Then he would grasp one's knee and say, "Lovely sounds darling." I got on well with David and his lovely wife Lucy, whom he called Bubula, but he should have been transparent with his bookings, not just to book his own favourite musicians.

Lansdowne was an approved studio on the Musicians' Union "fair list", although I never really found out why it existed for studios and for what purpose. Denis booked some session musicians direct, particularly those who were deemed "unreliable jazz guys" or inexperienced by the big session fixers. It upset some fixers; they didn't get their fixing fee! In 1960 the union set new rates of pay for musicians. A three-hour session in which 20 minutes of music can be recorded would pay Principals £6 (£123 today), Sub-Principals £5 12s (£116 today) and all others £5 5s (£109 today). (Source: Union history).

As a studio owner, Denis saw his role as an opportunity to give new talent a chance, enabling assistant engineers to record his sessions, particularly those who were ready to progress to engineering in their own right. I agreed with his approach because I knew the guys concerned would not mess up, having already assisted me on many sessions. The guys had a good "bedside manner" – one of many important criteria when I engaged them. I looked for people that were cheerful, creative, flexible, and able to understand what the client wanted even if it didn't often match with what was said, and above all no one with stars in their eyes.

Preston sought to make Lansdowne the best-equipped studio for sound recording in London, and indeed in Europe. A very big ask but it worked. Ultimately, clients came flooding across from IBC and elsewhere to see, and work, in this new astonishing studio – so different from the other few in London, designed from the ground up in the stripped-back shell of all Lansdowne House's basement and sub-basement areas.

The studio sounded good, but unfortunately it was located near the next door neighbour's property wall and sometimes at night they complained – it was futile, because I was asked to go round to see what the fuss was all about. It turned out that the neighbour, a Mr de Schwartzberg-Gunter, did not like a recording studio next door; at times he gave us ongoing grief. Twenty-five years later, I had a visit from this man's daughter, who was extremely pleasant. She apologised for her late father's behaviour all those years ago when she was a young child. She said, "He was always drunk and my mother divorced him, and I'm so sorry you had a bad time." Denis Preston was more scathing, copying the man's South African accent, while saying, *"that dreadful South African German Jew from next door!"* I can't repeat the rest, as it was somewhat anti-Semitic. I thought how considerate of his daughter to come to apologise after all those years.

Lansdowne's studio piano was a 1911 Steinway B in a solid rosewood case, which sounded great. Unfortunately, the studio was heated by electric heaters built into what would have been skirting boards, at ground level and below the acoustic treatment. There was no air conditioning! The amount of heat generated could be selected by switching on various heaters. The problem was that often the piano was against the studio wall when not in use and the dry heat started to cause the soundboard to crack slightly – a potential disaster for a piano. I had an answer to hang small clay pots of water suspended under the instrument the evaporation of which kept up the humidity and saved the soundboard from drying out and cracking. The pots needed topping up frequently – Joe would forget, so it was usually down to me. We later sold the Steinway (I persuaded Denis) to a percussion session musician Ray Cooper, a real gentleman, who later became Elton John's percussionist. We replaced it with a Bösendorfer 225 with a 92-key keyboard – a beautiful instrument with a big sound, very popular with the pianists and recorded well.

The work came in faster than management had anticipated and this caused a problem for Denis' projects as he found that he was competing with outside clients for time; a difficult position to be in. Joe was working his socks off – as he did – and more of a loner than ever. I was supposed to be his assistant and an engineer in my own right but this didn't happen at first, as Joe preferred to work alone. So I got on with checking in new tape from EMI and other mundane jobs. He had his so-called secret echo device all taped up. I asked Joe what it was, and "It's a secret" was the reply swiftly followed by the order "Don't touch it!" This was the usual stuff that I had heard at IBC, and I accepted that, but when he wasn't around I looked; it was an old HMV electric spring heater which Joe had modified for "spring echo". No secret then – it actually was not that good a sound but it worked of a sorts; it made a very good twanging sound when kicked! It sat on the control room floor and with loud monitoring in the control room the device picked it up and you could hear it when it fed back in the mix – not so clever! Joe had his multi-frequency band equaliser ("cooker") manufactured. A hangover from the IBC days, it was a Pultec-style multi frequency equaliser and once again he became very protective about these two pieces of kit.

On Joe's sessions, the control room was a solid mass of cables running all over the floor, plugged into this and that. God only knows what state the tape machine line-up was like – he wouldn't let me align the machines to the CCIR standard test tape! Up to his old tricks again! This worked fine if the stereo tapes were used in-house, but they weren't when they moved out with the client to be cut, usually at IBC, and there was never any line-up tone – zero! This made life difficult for the cutting engineer, having no line-up tones for level setting and azimuth adjustment.

I don't think that made engineer Ray Prickett at IBC happy, where later in the 60s, initially, much of Lansdowne's stereo output was cut to a master lacquer on a Lyrec stereo lathe. I insisted line up tones were at the head of every master tape that left the studio.

I recall one very amusing incident: a mono tape was sent to EMI Abbey Road for cutting a 45 rpm master lacquer. It was sent back with a note saying they were unable to cut it because the level on the tape was too high and there was too much distortion – that really showed the mentality of those Abbey Road cutting engineers – those guys in their white-coats again! Not only that, it was, and still is, (with digital CD post-production engineers) my opinion that a cutting engineer, or digital post production engineer, should not act as a frustrated recording engineer imposing their production will on the original material and producer, and think that they should "try to better the sound" – unless asked by some nervous producer who is not sure it was correct when listened to in the studio control room. It used to happen time and time again. Joe went mad with many expletives! Denis was most annoyed and said that never again will any material go to Abbey Road for master cutting.

Later, when Roger Cameron was at Advision in Bond Street we took all our mono masters for him to cut. I personally took master tapes (with line-up tones- crucial for the cutting engineer -) to Roger, and he cut them without any further ado or criticism on the Lyrec mono cutting lathe. On single releases, I asked Roger to cut them "hot", which required a deeper cut to prevent a stylus jumping out of the groove. He subsequently cut all the Dave Clark Five material hot!

Although Denis said he would never use "Shabby Road" (Abbey Road) again, there wasn't much choice, and the master was sent back to be cut – it was, but under what protestations I don't know. The disc was *Sing Little Birdie Sing* by Teddy Johnson and Pearl Carr, which EMI Abbey Road originally refused to cut. That says it all! It was voted number two in 1959 in the fourth Eurovision Song contest,

which was broadcast from Monaco. A hit in the UK, it climbed to
12. Joe won out again. Denis said never again would he go there
for recording any of the big orchestra recordings which were planned
later with Laurie Johnson, and he didn't – although he had his Record
Supervision Ltd contract with Columbia.

Denis asked me to enquire the cost of hiring Abbey Road No.
1 studio. The quote was £150 per hour in the middle 60s! Today,
using Bank of England retail price index – average inflation of 5.3%
over the years, it amounts to over £3000 per hour! An indignant
Denis commented, "I am not paying that to Sir Josef Lockwood to
contribute to the tarmacking of EMI Hayes drive!" How correct he
was. It later transpired that the management of Abbey Road were
opposed to any "outside engineers" coming to their studios to record
on behalf of a producer. It was rumoured that they wished to protect
their own engineers for fear of being shown up by the "outsider", who
might have proved to be a more creative engineer and who did not
understand the familiarities of the console, or who might technically
pick holes. Hence, I guess, the high hire cost to put off "outsiders"
booking the studio.

Denis asked me to collect some masters from Abbey Road, I was
invited for a cup of tea in the downstairs "canteen". One thing stuck
in my memory was the spoon in the sugar bowl was attached to the
counter by a piece of string! Today, they are wonderful, friendly
studios with great engineers. Zebra crossings come to mind and we
have a lot to thank the late Sir George Martin (Beatles producer and
many other artists at Abbey road when he was one of the EMI house
producers) and Geoff Emerick (EMI balance engineer who recorded
the Beatles) who, in my opinion, changed management attitudes at
Abbey Road. It is the norm today that engineers freely work around
many studios including those who are full-time employed by a studio
whose client may sometime work elsewhere and request the engineer
follow.

Chapter 6
1958 – 1960 Before Joe's Departure

Joe's work continued through to November '59, with his old IBC clients coming to Lansdowne as well as new clients. Wally Stott, an old client of IBC, recorded Diana Dors' *Swingin' Dors* with Joe at the controls. The record was released in red vinyl with a gate-fold sleeve – unusual then. This was the only complete album Dors recorded. Denis Preston continued his Record Supervision jazz recordings with Joe. I got to record some small jazz outfits. Ever since I knew and worked with Joe he never took a holiday. In June '59 he was due to go on holiday and hired a boat but he never made it because of his workload. The boat hire company were not happy about his no show and wanted their money for the hire of £19. 2. 6d being the remainder after the £10 pound deposit Joe had paid to the company "Maid Line". Lionel Stevens wrote to the company's lawyers. The matter was eventually settled for a less amount paid by Denis Preston. That was Denis for you – he took care of his people.

The first experimental session was in early 1959 to see how the studio floor performed acoustically and how the equipment performed technically. Naturally for Denis this involved the Chris Barber Band and it went well from the recording point of view. Chris was a purist and while there is nothing wrong in that, Joe had other ideas. He wanted to record the Chris Barber band in a more commercial sense and Chris did not like it at all. Chris and Denis didn't always get on – two strong personalities – and Chris expressed his feelings that Denis drank too much. He may well have done, but Denis was always on top of the production in hand. The band preferred not to be too tightly miked – no sounds right in your face. This was not Joe's forte, but he captured a not-too-tight a sound. I didn't know then the technical troubles that lay ahead for me. Trad jazz bands were notoriously temperamental about how they sounded and the Barber band was no different. I received a comment on this point from Richard Preston, about his father: *"Well that's why my dad and Chris fell out about the same thing. My dad thought you can't wrap jazz up in aspic in New Orleans in 1922 and particularly when it is played by a bunch of white English middle-class kids in England in the 1950s. It is the sort of music that moves on and what you are doing is fine and you're enthusiastically interpreting it but you can't have these arbitrary rules that if King Oliver didn't do it like that then we won't do it because, you're not King Oliver!"*

Some of the jazzers were not at all happy about the way Lansdowne went about recording them. To me, they seemed steeped in the past and no doubt would have been happy if they all sat under a horn recording to wax! I recorded Wally Fawkes (clarinet) and Sandy Brown (clarinet) for the Fawkes-Brown LP/EP recordings, and Wally said of his sound, "It was like being recorded at the end of a wind tunnel." He was referring to our use of echo. Johnny Dankworth, with whom I worked with Denis, disliked our recorded sound. Some of these guys couldn't get their heads around the changes. They were used to the old roomy one-mic pick-up recordings.

We recorded Cleo Laine (now Dame, and Johnny's wife), a wonderful singer with a great voice – she knew how to sing, and I greatly admired her. Johnny (known as John in the early years) Dankworth commented on her recorded voice quality to Denis complaining he did not like the echoey quality on the voice (our use of the echo chamber, again). Denis retorted, having had a few brandies and coffee, "I don't like the pissy sound you make on alto" (saxophone). That was Denis when in his cups, straight to the point, and Dankworth was an occasional Record Supervision artist in those early years as was Cleo a Record Supervision artist!

In 1960, I went on to record an album with John– the name of which I cannot recall, a marvellous player, and modified my recorded sound (it was before we moved control rooms) on his alto sax. I thought we worked well on that recording and, I think, on all his compositions - John's arranger was David Lindup - they were excellent scores and recorded well. He was an outstanding musician. Cleo was made a dame in 1997 and Johnny was knighted in 2006. I always enjoyed recording them in the '60s. Cleo had a wonderfully rich vocal sound, a tradition followed by her daughter Jacqui Dankworth. Alec Dankworth is a bassist. I had an enjoyable time working with John and Cleo in the early 60s with Preston producing. Their dry sense of humour was well matched. Johnny passed away in February 2010.

The studio acoustic design by Sandy Brown – a jazz clarinettist himself – was most suitable for that type of recording. Sandy was a BBC acoustician so he knew what it was about designing studio acoustics for BBC. It was not my place to comment at that time, and I held my diplomatic counsel. Personally, I thought the room was too live for its size, with a grey linoleum floor – it had an odd acoustic feel, but it worked for jazz. Ken Colyer, another Denis artist, wanted just one mic to cover the whole band and could prove difficult if he didn't get exactly what he was after – frankly, the end results were rotten. He was a traditional jazz purist but in my opinion had a dreadfully old-fashioned band sound – he wanted to replicate the style of 1920s New Orleans jazz. He had many followers so it was not my place to question, but personally it was a little dusty for me. I had an opportunity to record the band in 1960, with *This is Jazz* in the Lansdowne Jazz Series. Produced by Denis (SCX 3360), this was recorded in Lansdowne on 22, 23 and 25 August 1960. I trod very carefully and, with some trepidation, using the minimum of mics but a more modern sound with my fingers and everything else crossed! I had no grief from Ken. I understand today that the recording is hard to come by and would fetch a high price for the vinyl.

Joe's time at Lansdowne was most fertile. Even a *very* small sample of his work would include many albums: Lonnie Donegan, Chris Barber's Jazz Band, the Fairweather Brown All Stars, Acker Bilk's Paramount Jazz Band, Sonny Terry and Brownie Mcghee, and Marty Wilde. A further example would include The Lansdowne Orchestra & The George Mitchell Choir, and their *The Blue and The Grey* album, which served to highlight one problem with some of these first Lansdowne "experiments" in stereo. The album was badly out of phase, and it proved a nightmare to get an acceptable mono result from the recording. The master Jazz "stereo" tapes we received for Denis from the USA, for example Booty Wood, were in what we called ping-pong stereo – left-channel, right-channel and little in the middle. At least we never had phase problems with these when reduced (an old term now called collapsed) to mono.

Later, to my dismay, I discovered that many of the early stereo recordings exhibited phasing problems – sometimes more, sometimes less, depending on the choice of microphones that Joe used. Some of the dynamic microphones were out of phase to the Neumann condenser microphones, and on sessions he also used a number of dynamic mics. The question arises as to why did Joe did not "get it". He was technical but not "technical technical". Now I understood what the techs at IBC had to put up with.

When I had an opportunity to record a session I never knew what the line-up state of the recorders would be, and living with Joe's cables going everywhere was a nightmare. Before I could record it took some time to check everything technically. Joe's EMI tape wastage was, and I don't exaggerate, colossal! He would use half a reel, then discard it, or during editing he'd leave the out-takes on the floor, which resulted in my clearing up. At this time, he was becoming more and more dependent on Preludin, and would flare up at the slightest thing and go into a sulk – I ignored it, and he would then eventually be back to his normal old self. He sometimes worked all night on his own productions, which he called RGM Sound. After Joe left Lansdowne, he recorded from his flat near Lansdowne – before Holloway road. Unbeknown to Preston or to me, Joe had already set up his own company, RGM Sound, at his flat. Joe worked nights at Lansdowne to further his own productions. He could be quite cagey about this when I asked him about doing this behind Denis' back.

I wondered if it was with the tacit approval of Denis. If Denis knew, he let it ride because I thought he wished not to upset Joe (they got on so well together), who frequently ran on a short fuse. So it was, just Joe and I, and much work for me trying to keep on top of matters technically and trying to get Joe to use all the reel of tape; there were half-empty tape spools everywhere!

One day when I arrived at the studio I noticed parked right outside Lansdowne House entrance (much less traffic and no parking meters in those days) a new white Sunbeam Rapier convertible – it looked immaculate. I found out it belonged to Joe, and it had been bought by the company as part of a contractual deal that Joe had with Denis. Before the studio was really up and running we made frequent trips to the Bush (Shepherds Bush) to eat and, sometimes, to an old wireless surplus shop in Edgware Road – Smiths, I think it was called – because Joe was attracted to surplus radio bits and pieces. On one of those occasions the car was looking a bit forlorn, so I asked Joe if he ever cleaned it, and the answer was no. The car was looking progressively worse and I thought, how could he let this happen to a brand new car? I said to Joe, "Why don't you look after it, how about checking the oil and water?" His answer was, "I don't bother". What a pity. Whether he did eventually I am unable to answer and, after that he left it parked outside his flat and walked to the studio. I suspect that he was so single-mindedly focused on his recording work that cleaning his car, or cleaning the studio of discarded recording tape or even taking a holiday was of no consequence to him. It was his work that mattered; all at the expense of everyday life. The continuing workload for Joe was demanding, and clients were attracted by a studio built from the floor up with all new equipment. EMI only usually built valve (tube) consoles for their own studios (EMI Reds – much sought after today – Mark *Knopfler's* British Grove Studios has one in immaculate condition), and the Lansdowne version of the console's work surface layout was suggested by Joe. At that time, I would usually work a nine-to-six day, five-days a week and, indeed, was happy to work longer hours. Although I was supposed to be an engineer from the outset, it didn't work out. Joe was very jealous of his old clients from IBC, and also the new Lansdowne clients, and was paranoid about anyone else getting the opportunity to work with them – I didn't get a look-in! Frankly, the small control room was a complete shambles and a mass of wires going to God knows where, and discarded partially used reels of tape; it was impossible to do anything and, of course, they were *his* secrets. His talent, as always, shone like a dazzling beacon in the night. Despite the enormous amount of work, Joe insisted on doing it all – but in the end it got to him. I was often sent to the chemists to buy more Preludin, which I saw him put in his mouth like sweets – he regularly worked all night on clients' material, tweaking this and that, or his own material, and was still there when I arrived the following morning. If there were no sessions on until later in the day, or the studio had a break, Joe would go home for a few hours' rest. It got worse through the first few months of 1959.

Then not surprisingly, it happened, and the mucky stuff hit the fan – big time! It was on 4th November 1959. Joe was on another stereo recording session (in those early days Preston's recordings were usually mono) – this one was with Denis, producing his artist

Kenny Graham's Afro Cuban Band, which Joe had recorded before for Denis at IBC. Kenny Graham was a saxophonist, composer and arranger. Joe kept pestering Denis during the session, to listen to and record some of the songs he had written for Mike Preston, including *Put A Ring On Her Finger*. He wanted Denis to record Mike. Denis lost his patience, "For Christ's sake Joe we are doing a session. This is important, we'll talk about your songs later". Unfortunately Joe also lost his temper and patience, and walked out of the session, leaving it engineer-less! What a stupid thing to do to all people. Denis didn't take kindly to Joe's attitude. It was the straw that broke the camel's back. Denis called a tea break. I was working quietly in the unfinished upstairs "stereo" control room trying as usual to sort out Joe's mess of tapes and unpack a new delivery of virgin tape from EMI when I saw Joe come up the stairs from the studio and stomp out of the main studio entrance door. The next thing I know, Denis called me downstairs to a wire-tangled mess of a very hot (over 100 valves dissipating heat), airless cigarette smoke shrouded control room. Puffing heavily again on his full-strength *Gauloise* fag, with a shot glass of whisky in his left hand he asked me, "Can you take over the session?"

"Er, yes," I said, with some trepidation. I knew how Joe worked, but first had to untangle his set-up to understand what he was doing technically. I asked Denis for an extended tea break to give me time to sort out the tangled mess, and he agreed. During the break, I resolved most of the mess by following cables back to source, and the session was successfully completed on time – just! I really sweated on that session, as it was a large band costing a lot of money in musicians' fees. The recording levels Joe had set meant that everything was in the red and *I mean in the red* big time – the band was loud with much percussion – I really struggled.

The next day, Joe arrived as if nothing had happened the day before – no surprise there, the pills had worn off. He was summoned to Denis Preston's office and was dismissed on the spot. It was an immediate and abrupt end to an immensely fertile and creative relationship. Joe had made a crass mistake in pestering Denis on one of Denis' sessions. His contract was terminated with immediate effect and I never saw Joe again after that. It was a great shame as I owed so much to him. I missed that man who was so kind to me – a true human being, very badly misread by people who were jealous of his immense talent. He was the master of his own undoing – he had pushed Denis too far. Joe left his trademark legacy of hits for the studios: Chris Barbers *Petite Fleur* recorded at IBC in 1956, charted #4 February 59 for 18 weeks – how ironic! Emile Ford and The Checkmates' *What Do You Want To Make Those Eyes At Me For,* #1 for 17 weeks in October 59, I was on that session. Marty Wilde's *Sea of Love,* September 59, charted #4 for 12 weeks, Mike Preston's *Mr Blue* in October 59, charted #9 for seven weeks. Joe *was* the pioneer of many recording techniques used for popular music recording today. In his lifetime, and even today, he hasn't had the deserved recognition for the contribution he made to our industry.

The day after Joe was sacked I was summoned over the intercom system to Denis Preston's office (affectionately known as DP). The intercom system covered all office areas and the control room – it was such a pain in the arse, especially when Lionel Stevens (an FCA accountant, known as LGS – Lionel Gordon Stevens) had had a few large gins! Zzzzt it went... my turn to be summoned to DP's office. On entering, I saw that Denis's partner, Lionel was there – they shared the same office. "Joe's gone. Not coming back. His contract is terminated," they said, with no mentioned about the previous day's events, simply, "Can you run the studio for us? There will be more money." Well, that took the wind out of my sails, how could I follow an act like dear old Joe?

A 22-year old given that amount of responsibility in the blink of an eye – I was so taken aback, my answer was, "Er, um, yes, oh"!

They replied, "We'll give you a probationary period of… let's say one month and, if satisfactory, a contract and more money." But I was not yet told how much more money. "Oh well Kerridge, get on with it," I thought. An opportunity not to be missed!

Then the realisation hit me – just little old me and all those clients. I need help, well at least another engineer had to be recruited, and fast, I thought. I didn't how much of a huge technical mess there was to sort out.

Then there was the matter of Joe's unfinished business. The very next day, 5th November, Laurie Johnson was in the studio to complete the last session of his *A Brass Band Swinging*. Oh no, I thought, where do I begin? I didn't have Joe's set-up from the previous sessions, on 6th and 9th October, so I took a chance for the set up and guessed how Joe would have done it on the previous sessions, as I had not been present. Fortunately Laurie arrived early "to case the joint" and advised me on the set up which required some small changes. A result! I discovered later, as with the Kenny Graham session, phasing problems were present. Later on playback I could hear them this time, but there was nothing I could do until I'd investigated – all the sessions were in stereo!

The Kenny Graham's Afro-Cubists stereo sessions lasted the whole day, two sessions at three hours each: 10.00 to 13.00; 14:00 to 17:00. In the lunch break, the musos went to the pub for some liquid refreshment. I edited the sessions after the gig.

Recording sessions with musicians then and today (apart from group band recordings) were usually in three-hour slots: 10:00 to 13:00, 14:00 to 17:00 (or 14:30 or 17:30), and then 19:00 to 22:00. The breaks gave musicians time to get to another studio if they weren't engaged in the same studio at an all-day recording. In later years, and in some instances for film scoring, the sessions increased to four hours each.

I discovered later this session was not the only one in stereo that had a phase problem. Denis' material was normally released in mono and, in those days, we used a term called reducing stereo to mono; later we called it collapsing to mono – in the multi-track days, mixing to either format. I played back the tape through the console to reduce it to mono – at least, that was the idea for release in mono – but disaster! The whole band sound disappeared down a "tube" with the echo prominent and, to make matters worse, some of the band sections were more present than others! Bloody hell, I thought, we have major phase problems here!

Initially, I thought the primary problem was that the inputs to the stereo machine on the original session were 180 degrees out of phase. This term is not strictly correct because in a complex audio signal not all frequencies will be (180°) out of phase. In non-tech terms, the audio cancelled out when the two tracks were combined; something I should have picked up with my ears on the original session from the monitors but didn't as everything was done in somewhat a hurry because of the delays on the previous morning's session. Without going into complex technical detail, I sorted it out as much as I could – what a rotten sound because of those tech errors but we had to live with it. I had to find a solution very quickly. Where exactly in the signal chain was the phase problem? There was also the echo chamber to equip with a decent speaker and two omni-directional mics spaced apart for left-channel and right-channel reverb returns, not to mention fixing the badly fitting door and painting the walls and floor in gloss paint. One of the echo chamber walls was a simple dividing wall and resonated when thumped – not good for a chamber. The answer was to build another plastered brick wall in front, with the gap behind filled with sand – it didn't move or resonate!

If I was going to run the technical side of the studio, I was adamant that I required more staff to cope with the workload. I put the request to Denis that this should happen, and it did. Dave Siddle joined me at Lansdowne in December 1959 followed by Vic Keary nine months later in September 1960 and we set about sharing the workload – the book was full for session work, and the phone rang constantly. After Joe's dismissal – and Dave commencing – Denis employed an engineer called Dick Lazenby but he was not a memorable engineer and left the company in May 1960. Denis also employed another engineer in March '61, Bill Johnson, who claimed he was Denis's assistant – not true – and also an unmemorable engineer. Johnson was sacked for showing 16mm blue movies to all and sundry at night when the studio was not working. Although Denis was liberal minded this was not tolerated by him once he discovered what was going on. Until this time it was Preston and Stevens who were the hirers and firers. It was after the dismissals that Denis left it to me to judge who we employed for engineering, studio bookings and receptionist. With engineers Denis gave them "the once over" to see if their personality and sense of humour was suitable to work with him.

Dave and Vic were competent, conscientious balance engineers and Vic was technically knowledgeable about valve gear. Preston had another company (amongst others), Nonesuch Records, to record the spoken word. Vic and I shared the work, as the individual interviews were recorded remotely from the studio. We didn't need much kit to record; a transportable TR51, one mic and monitoring on the machine's internal speaker. I recall Denis interviewing Noble Prize winning writer Aldous Huxley at his flat, and I positioned the gear in an adjoining room – only one mic required.

Half way through the second reel of tape, I fell asleep and woke up to find the machine in stop! I quickly changed reels, feeling an absolute fool, and thought I'd be in trouble, but no, thank goodness, because when it came to the editing Denis didn't notice the lack of continuity – phew! I went on to record Lord Birkett, renowned legal mind and alternate British judge sitting during the Nuremberg WWII Trials.

Then in 1961, we had in the studio a dramatization of the trial of "Lady Chatterley's Lover". This focused on the book's trial for obscenity and was a long, complex production starring Maurice Denham, Michael Hordern, John le Mesurier and others. It was a two-12″ LP set, which I had cut at Pye Studios. However, on taking a test cut back to Lansdowne, the bass end of the voices had disappeared almost entirely and the recording sounded thin. The problem lay with the Pye Studios' cutting room acoustics, which featured a Lockwood cabinet with a Tannoy monitor speaker located in a corner, so the bass ran "down the walls" and was exaggerated. I have forgotten who cut it, but clearly remember the heavy bass in that room that resulted in a bass cut for the mastering. I thought we had a problem with monitoring at Lansdowne, but thankfully not. Pye insisted the room was OK; we agreed to disagree. It was an early lesson for me – to make sure room acoustics were well designed for monitoring. When I returned for the re-cut, I had it mastered flat without any cutting room EQ. Job done. Other recordings for Nonesuch were of Richmal Crompton (author of the Just William stories) and an unmemorable recording at Chelmsford Cathedral using hired-in gear from where I cannot remember. What I do remember is that it was damn cold in that Cathedral!

Vic recorded Bertrand Russell, the philosopher and Nobel Prize winner in literature, at Portmerion, being interviewed by a producer called John Chandos, who produced most of the spoken word series. Vic edited most of the material in our Studio Two, which was a small room at the time. We used Scotch Boy 102 recording tape, which had a strong solvent smell – the room was poorly ventilated. Vic said, "I think I got stoned on the rather pleasant smell!" In the later 60s, another person involved in the spoken word series was Dame

Edith Evans, with John Mackswith engineering. John takes up the story: *"Another grandiose DP production, the title of which escapes me. Denis invited me to a pre-production meeting, to discuss arrangements, layout, rig and finally his desire for Dame Edith to narrate the storyline. Of course I found it quite astonishing that Denis had such a powerful command over the entertainment industry. Dates and schedules were arranged. On the day of recording Denis suggested Dame Edith be on the studio floor, not in either of the separation rooms. Set in the corner adjacent to the booth were table, comfy chair and screens to assist separation. Whist adjusting DE's mic she exclaimed, "Young man, don't point that thing at me, take it away!" Shocked and aghast, I returned to Denis, informing him that DE thought the mic an intrusion, to which he replied, "Do something..." Innovation, quick, help, come up with a solution. Eureka... discreetly place a KM54 in the sleeve of the assistant engineer and advise DE that he will turn the pages of script silently, to relieve her of the task. Eureka – problem solved, QED!*

As the work was increasing, I met with Denis and Lionel and asked for more staff, namely a studio attendant, cleaner come tea server for the musicians' breaks and to help with the studio set-ups, as well as a bookings person to handle all the calls for bookings. Preston's secretary could no longer do this, as she was busy with Dennis' increasing workload. This was all approved. At times, as an accountant, Lionel Stevens could become quite disconnected from the realities of the studios' staffing needs. Denis and I dug our heels in. I knew I was on probation, but needed to sort out the technical side of the studio and quickly, while also handling the recording sessions. Denis supported me. It should not be forgotten that the work was not only for Denis, which involved a heck of a lot of jazz recordings, but other well-paying clients too! Naturally DP, as majority co-owner of Lansdowne, enjoyed a good discount on studio time for his Record Supervision projects. The discounting rattled me when DP occupied the studio for most of the week on various projects because

LGS would complain about lack of profit therefore no money for investment – and *he* was an accountant!

I told Denis and Lionel that I required time technically to go through all the equipment signal chain, and sort out the prime trouble – phase problems – to make sure everything in the studio was phased correctly, and that meant the microphones as well. To put this into perspective, I knew the Neumann's were OK (what about the dynamic mics?) but it could be the microphone cable, wiring connections or the cables to the microphone amplifiers or the problem could be in any other parts of the signal chain. EMI had installed miniature connectors of Joe's choosing that were small and prissy, always falling apart, and difficult to plug in and take out. I found many of them not wired correctly – i.e. cable inner red to a specific pin, with the black and screen also to specific pins – not so here; it was a case of random wiring. Other jobs included installing the industry-standard Cannon connectors (Lionel wanted to know why), have new mounting plates et cetera made up, and the connectors wired to the correct standard, which I did, and rewired them myself. Meanwhile, the studio could only accept limited work for three weeks while the rewiring was taking place.

Chapter 7
Practical Issues of Sound in Studios

The universal, everyday experience of sound

Every day of our lives, we are exposed to sound in our environment in one form or another, even before birth. Though we generally take sound for granted, we naturally become more aware of it when listening to music, whether live at a concert or broadcast on radio or television, recorded on CD or vinyl, or downloaded on our computers and iPods, or when it is accompanying film – in cinemas, on disc, or in television programmes and advertising.

There is an absolute need to record (or evaluate the recording process) in an acoustically-controlled environment. Recorded sound, particularly music entertainment, is almost universally available and, in order to achieve the best possible results, and for it to translate into such widely varying situations and formats, the recording engineer must record in a controlled environment. For those professionals reading this, that is taken for granted – if you are not hearing an accurate representation of what your music sounds like during recording, then there is little chance of producing a high quality end result. I have had the none too pleasant opportunity to record in a number of studios over the years where the acoustic-architecture has been appalling, where a controlled and balanced room frequency response did not exist, and the client was not at all happy with the

131

results – *"It doesn't sound like that when I play it at home. What did you do?"*

How many times have you heard someone remark: *"I went into this or that building/concert hall/room and it sounded odd when I spoke"* or *"I could not clearly hear the music/the speaker"* or *"There was too much echo"* et cetera. The acoustic in which we hear sound has an enormous bearing. In some concert halls and other large enclosed spaces that are not particularly well designed with acoustics for listening in mind, a common defect can be excessive reverberation – the slow decay of speech or music (a long reverberation time) due to the repeated reflections occurring at the hard, relatively smooth surfaces of walls, floors and ceilings before the energy is finally dissipated.

The need for a balanced approach to acoustic control and the avoidance of excessive high-frequency absorption

An extreme and, unfortunately, all too common solution would be to cover the all of the surfaces with sound absorbent materials. However, this can lead to over-correction of the reverberation time, particularly at higher frequencies, producing a deadening effect that can be most uncomfortable for performers of either music or the spoken word, as well as for the listener. Many older studios and control rooms were treated in this way, by acoustic consultants with an out-dated mind-set. The higher frequencies are vital, to preserve the harmonics that define the very timbre and quality of the sound.

Sabine and the beginnings of architectural-acoustic investigation

As early as 1900, an American physicist (Wallace Clement Sabine, 1868-1919) became a pioneer in architectural acoustics, devising experiments to investigate the impact of absorption on reverberation time. He was ultimately able to define the behaviour formally, and Sabine's formula still helps us to determine important characteristics for gauging the acoustical quality of a room, providing the ability to predict how "wet" or "dry" a room is likely to be. Sabine derived an expression for the duration T (time) of the residual sound to decay below the audible intensity, - the time for the level of energy to decrease 60 dB also depends on the volume of the room, it is a complex subject - and laid down three simple rules that must be followed if satisfactory results are to be obtained. 1) The sound heard must be loud enough; 2) The quality of the sound must remain unaltered – that is to say the relative intensities of the components must be preserved, and 3) The successive sounds of speech or music must remain distinct, i.e. there must be no confusion due to overlapping of syllables – whether sung or spoken. These are basic principles that hold true today.

The first auditorium designed by Sabine, applying his new knowledge in acoustic architecture, was the new Boston Music Hall (Symphony Hall), formally opened on October 15, 1900 and still considered one of the finest concert halls in the world. The unit of sound absorption, Sabin, was named after him.

Thus a new branch of physics – architectural-acoustics – was born. It was on 29th October 1898, at Harvard University, that he established the precise nature of the relationship between these quantities and placed the subject on a scientific footing – long before the invention of the recording studio and the new breed of architects and designers that was to emerge into a new world of "recording architecture".

The Precedence Effect, the control of standing waves and the importance of other acoustic parameters

Reverberation time is no longer considered to be the only acoustic parameter that must be addressed in a recording studio. It is vital to consider the Precedence Effect, the psycho-acoustic phenomenon whereby the brain fails to separately distinguish a reflected sound arriving within a few milliseconds of the direct sound, for example from the monitor speaker in a control room, and effectively perceiving a single and misleadingly-enhanced sound. The best control rooms I have ever worked in are those in which the sound arriving at the ear is not coloured by the room or by multiple standing waves, or by the monitoring system itself, which must deliver a faithful reproduction of the original instruments. Achieving such lack of colouration gives the engineer the chance to equalise the various sound components, and to be creative in the final production without the worry of "what will it sound like when I play it at home or in the car?" The above, of course, holds equally good for post-production mixing rooms for film and television.

The significance of wavelength variation over the audio spectrum

The generally accepted frequency sound spectrum for audio is 20Hz to 20Khz (harmonics are a different matter), which in itself poses a potential problem. The wavelength at 20Hz is a huge 17.2 metres and at 20Khz just 17.2 mm. For large halls, such long low-frequency wavelengths may be easily accommodated, but not so in many recording studios. Especially in control rooms, the studio designer is often constrained by spatial limitations that generate complex and difficult design criteria. Mathematically, one could not squeeze a quart into a pint pot. Primarily in the 70s, this was overcome by large amounts of deep absorption and bass trapping that engulfed a large part of the room – not at all satisfactory if space is at a premium – and created an unbalanced, overly-deadened acoustic. I visited a number of studios in Los Angeles in the 70s and early 80s, and saw and heard this for myself. This was studio acoustic design by "a flavour of the month" designer. The same style ultimately came to the UK and a number of studios were built in the same fashion – so much space was lost. The monitoring loudspeaker's frequency response was corrected by White one-third octave equalisers – phase shift abounded!

The art of tuned, low-frequency membrane absorption, pioneered at the BBC with Sandy Brown, appeared to be lost! Not at Lansdowne in the early 60s. The effective use of diffusion rarely made an appearance until the 90s.

The importance of combining architectural/acoustic disciplines in a coherent design strategy and the value of experience

I have met them all – from architects not understanding the technical and acoustic needs of studios, to acousticians not grasping the ergonomic and three-dimensional practicalities of architectural space, as well as architects caring only for abstract visual statements, or acousticians preferring to bury their heads in data rather than looking and listening to what is actually going on, and with the builders grinning with a reassuring *"Don't worry, Guv', I've built studios before."*

The one designer I had previously encountered in my professional career who actually seemed to grasp the issues, (a career that started in the early years of this still relatively young industry), was acoustic-engineer/architect (and significantly, clarinettist), Sandy Brown. As already mentioned he was responsible for the original design of Lansdowne Studios, where I cut my sound-engineering teeth with the likes of The Dave Clark Five, big band and jazz groups and, as mentioned previously, working alongside the legendary maverick, Joe Meek.

In the early 70s there was one other designer of similar standing: Eddie Veale and Veale Associates. When I asked Eddie to measure the studio acoustic reverberation times at Lansdowne, we discovered that some of Sandy's membrane absorbers were not doing their job – not made to specification, and therefore badly made. Was it another case of, *"Don't worry guv it'll be alright on the night"*? The matter was soon corrected, by installing new absorbers. On making the new measurements, the reverb times were good, with the expected reverb time rising at the lower frequencies – however, they were within acceptable limits.

Decades later, I was introduced to Recording Architecture (RA), founded by Roger D'Arcy and Hugh Flynn, by one of Lansdowne's engineers, Jonathan Miller. I was very impressed with their philosophy, in which they recognized the need to bring together the aesthetic, ergonomic and technical requirements of a recording studio into a single, cohesive architectural/acoustic design solution. I used them to rebuild Lansdowne's control room in 1979, from the shell up, to accommodate surround sound. The build finishes were impeccable. RA used a very experienced building firm who really understood the acoustic requirements. We achieved a stupendous sound, together with the new ATC monitors. When we acquired CTS, RA again worked their magic with modifications to studio one control room and studio floor (a "fashion-built" control room, with much space lost by trapping) and a total rebuild of studio two, the all-digital facility of which I was very proud.

Technical Background

Phase response and phase shift is a complex subject, so I will try to explain it in simple terms for the not so technical. So, how did we check for phase problems with the mics or rather the input to the desk? Simple, as a reference, use a mic known to be wired correctly and plugged in one of the channel inputs to a fader – the mic to be checked is plugged into another channel input on an adjacent fader. We set the mic amp gains to be equal (having measured that first), place the mics closely together and ask someone to make a constant *aahhhh* into them, without varying the frequency of their voice. I raised the mic's fader to be checked and, if the *aahhhs* cancel each other, we have a problem that is easy to sort. A long, slow but effective process. Neumann had a phase checker, but I preferred the voice method. It was much cheaper, and more realistic than a piece of tech kit clicking into a microphone – the real world rules!

I also discovered phase problems in the desk by virtue of all the transformers used in the chain with the channel EQs set flat. Using clean generated sine waves, I could measure the phase angles on a dual-beam oscilloscope, comparing the clean sine wave (direct input to the scope) with the one passing through the signal chain at different audio frequencies. It was interesting to see the results. The phase differences were not all that great – a little over 10 degrees – but variations to be expected with the applied frequencies over the audio spectrum 20Hz to 20kHz $>5°$ leading at 20Hz and $>10°$ at 20kHz lagging (relative to 1kHz). The biggest problem, I discovered, was transient response when I measured with square waves. A square wave contains all of the harmonics available in the frequency response being considered. The accuracy of the shape of the square wave tells us how well the electronics perform. The angle of the leading edge of the waveform tells us about the transient response: a nearly vertical line means "very fast", which is what we need for accurate sound reproduction. The angle of the top of the square wave indicates how good the low frequency response is: totally horizontal wave top indicates a very good LF response (usually well below 20Hz); a sloping waveform top indicates -3dBu roll-off at or above 20Hz. Since transformers were used in the mixing console design there was some "ringing". I cured this problem by experimenting with different capacitors on the input circuits. It worked very well and there was a much cleaner sound – a much improved transient response.

I was always particular about bass guitar sounds. I used a specially designed transformer, which matched, or bridged the output impedance of the bass guitar pick-up and then matched to the input impedance of the mixing desk – usually the mic input/line input. The lead from the bass guitar to the transformer was a low impedance loss lead whose inductance/resistance was low, mostly a coaxial lead – inductance and resistance will tend to minimise cable phase or time delays. This preserved the low full frequency range of the instrument i.e low phase shift and no (relatively) high frequency losses. The transformer was also designed to handle very low frequencies and to have a very small phase shift at the frequencies it had to handle – something like >5° leading at 20Hz and >10° lagging at 20KhZ relative to 1Khz.

So what is transient response? It is the first shock of an instrument note, and therefore it is the ability of the audio equipment to authentically reproduce a sudden waveform – a transient. It is a short-duration, high-level, high sonic energy peak, particularly generated by the percussion family of instruments including piano, snare drums, handclaps, tambourines, triangles, cymbals and other instruments such as acoustic guitars. In fact, as I discovered by experimentation, it was important to many other instruments of the orchestra as well.

The audio equipment must have the ability to reproduce the steep wave-front or transient faithfully. It makes an enormous difference to the sound, which becomes cleaner, clearer, and with more impact (punch) and tremendous energy. In a recording console it is of prime importance. One must get the transient response correct as early as possible in the recording chain with minimum phase shift through the signal path.

Let us consider some physics:

(i) A square wave contains all harmonics in the bandwidth being considered;

(ii) The shape of the square describes the bandwidth (the frequency range) and the transient response of the circuit(s):

(a) leading edge = rise time (Transient response);

(b) flat part = frequency response (horizontal = flat; sloping = band limited)

(c) falling edge = Transient release time.

"Ringing" describes artefacts that appear as spurious signals near sharp transitions in a signal. Visually, they appear as bands or "ghosts" near edges; audibly, they appear as "echoes" near transients, particularly with sounds from percussion instruments.

We conducted these measurements at Lansdowne in 1960, 56 years ago, and to the best of my knowledge and belief they had never been considered or experimented with earlier in the same context.

There are some who will say it is not as important as the tape machine; that, by design, it will cause transient loss. To a certain extent this is probably correct – but not always so. Well-designed tape machine input electronics will give very good transient results; in other words a very faithful reproduction of that audio recorded onto it. Tony Waldron, of CADAC, commented: *"The transient response of the initial input amplifier (mic or line) is the one of the two most important parameters for the collection of an accurate waveform – frequency response being the second. Additional processing/recording electronics after the initial conversion circuits produces much less damage to the signal. The way that I look at it is: the microphone translates a three dimensional soundscape into a two dimensional electronic signal. The input amplifier **must** have the widest frequency response and best transient response possible for accurate reproduction"* I absolutely agree with that opinion; it supports the experiments I did later with Clive Green of CADAC, when he was designing the first CADAC console for Lansdowne.

The EMI tape machine's electronic inputs had a pretty good transient response, but when we took delivery of an Ampex 300-4SSC (four-track console version with Sel-Sync) ½" all-valve four track tape machine running at 15" (38cm/s) or 30" (76cm/s) per second, costing £2,235 (today £43,013), the results were unbelievable at a recording speed of 30″ per second (76 cm/s). The bandwidth at that speed was pretty much flat from 20Hz to 27kHz – OK, I hear you say, the ear can't hear up to 27kHz and maybe so (it's also a function of age) *BUT it is important* to faithfully reproduce the harmonic structure of a note, which in some instances can go very high indeed. For example, the harmonics of muted trumpet can extend to 80kHz; violin and oboe, to above 40kHz [Source: James Boyk, CIT (California Institute of Technology)]; a crash cymbal can go up to beyond 50kHz, although the harmonics diminish the further up the harmonic scale they extend. It is seen by this small example that certain sources may still pose a problem to the audio equipment designer and recording engineer.

A small example for clarity, **assuming no channel equalisation used**:

> Let's take a violin A string (an A above middle C). It will vibrate at 440Hz (Piano A). This is the fundamental frequency or first harmonic and the second harmonic then will be at 880Hz. The second harmonic is twice as fast, therefore this is an A an octave higher and so on – the higher the octave above the fundamental, the less proportionate harmonic energy contained in the note. The ratios always remain the same. So we can end up with harmonics with an energy reaching up to 40 kHz or beyond, although much diminished, such as strings, woodwind, brass and percussion etc. as described above. The phase corrections and the subsequent transient response improvement made to the EMI console and the wide-band frequency recording at 30″ per second on the Ampex, gave Lansdowne a sonic edge over its competitors, especially its string and percussion sounds.

Because of the sonic quality of our equipment and applied recording techniques, Lansdowne was frequently called **The House**

of Shattering Glass or **The Holland Park Hit Factory.** Laurie Johnson: "a favourite of those seeking the very best recording facilities," [*Noises in the Head* pp.75, Bank House Books, 2003]

Joe, having sadly departed with some acrimony, should be given credit for what he did for Lansdowne. He was responsible for putting together for Denis the whole of the studio project, with financial oversight by Denis and Lionel, and for giving me so much inspiration to be creative in recording. I shall always be grateful for that. Vic Keary: *"Joe had left Lansdowne when I joined. I came for an interview a month or two before with Denis Preston and was told there would probably be a vacancy shortly, but I never actually met Joe. Although I heard him on the phone at EMI crying nonstop the day before he died. An hour later I passed him standing by his car in Holloway Road looking really dejected. Thought I should stop but decided against it as he didn't know me."*

How very sad.

Analog Tape Machines at IBC.

Cilla Black and Dave Heelis LRS Old Control Room.

Joe Meek Compressor Front.

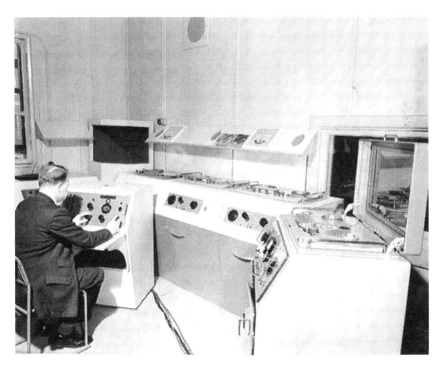

IBC Assembly Room Mid 50's.

Joe Meek with the EMI L12 portable recorder Paignton (Devon) sea front March 1956.

Paignton Devon with the EMI L12 portable recorder March 1956.

Royal Signals, Army Troop 1957.

Console 1958 from Newman Street in the half finished Control Room.

Joe Meek at the EMI Console 1959 with the usual mess of cables.

With Denis Preston in 1960's.

1960 Control Room At EMI (Meek) Console 6x4.

BBC PPM 60's Style.

With the DC5 and Lenny in Shot 1963.

1964-09-30 Raymonde session.

Johnny Pearson Conducting Ariola Studios 1966.

Ak and Syd Dale 1968.

With Robin Philips In the Ariola Control Room 1968.

KPM session at Ariola June 1969 wth Jim Lawless.

With Robin Philips Trixie 1969.

Mixing at Ariola Studios 1969.

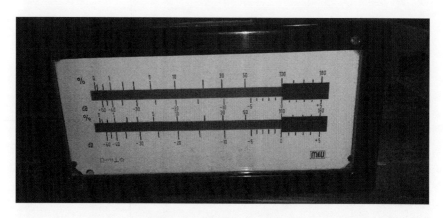

M&W Light Beam Meter.

Chapter 8
Life after Joe left Lansdowne

After Joe's departure the studio became busier and busier Dave taking on a substantial workload of DP productions. He was a competent balance engineer but not that technical. Over time we became very good friends outside of the studio. I was privileged to be asked if I would be godfather to his newborn daughter. I accepted, it was heart wrenching when at a very young age she died. I went to the funeral it upset me to see this small white coffin at the church: an event I have never forgotten.

I spent a lot of time sorting out the phase problems but it served me well to prove to Denis and Lionel that I was capable. The Kenny Graham sessions were resolved to an almost satisfactory conclusion, and after I had served my probationary period I was given a five-year contract and a salary increase from £15 per week (£312 today) to £20 per week (£415 today).

The appointment gave me the opportunities I needed to get the studio further improved on the technical front, and I was eventually appointed a director in May 1962 and was gifted a small amount of company shares.

Now being in charge of the studio, my third session (after the Kenny Graham technical disaster) was to record Diz Disley, a guitar player in the style of Django Reinhardt, on an early Sunday evening session. I arrived fifteen minutes before the session – small band, not a big set up, so a no-brainer – wrong! I got one hell of a rollicking from Denis, who went on,

"You must never be late for sessions."

"I wasn't," I protested.

"You have to set up and we'll be late in starting, don't let it happen again!"

Suitably admonished, I never did let it happen again. The recording went well, except I really pushed the console level almost to a fuzz – nobody noticed – musically, it was a very busy band.

In the early 1960s, I spent much of my time recording Denis Preston's work especially on a very time-consuming project with a guitar player (and various other instruments in the guitar family including Ukulele) Wout Steenhuis, who used to play in the Dutch Swing College Band. I spent hours in the studio control room with Wout, an excellent player and a real gent, recording Hawaiian music using the composite master technique that Joe had used at IBC and now me at Lansdowne. Oh, how much easier life would be to have multitrack! Wout Steenhuis, guitarist: *"I made 22 LPs at Lansdowne from 1963 onwards for EMI. I don't think we'll ever forget the first LP – multi-tracked between two mono and one stereo TR90 tape machines. How we had the nerve I'll never know, but on various instruments – for the 12 tracks – I played something like 138 times. This may seem unremarkable now, but at the time that album was a complete breakthrough. And it is still selling today".*

With careful machine line-up, and because the console electronics were much improved, there was little sonic degradation.

A technical problem I encountered was that of reverberation. To be more creative I wanted to purchase an EMT 140 (ElektroMess-Technik) reverberation plate to use as another source of reverb in addition to the chamber. It was deemed too expensive – Stevens's cash flow ruled. Wout suggested a guy from Denmark, a Mr Foss (I never did discover his Christian name) who made mono steel mesh plates.

They were very reasonably priced so Stevens agreed to purchase one. To house it we had to make an open wooden frame. Mr Foss arrived with a rolled up piece of expanded sheet metal, he unrolled it, hung it up in the frame, installed small transducer to excite the metal and a modified Ortofon record pick-up for the reverb return and the associated electronics. It was a primitive device of *fixed* reverberation of about 2.5 seconds, well at least it was better than Meek's spring echo. It was damped to achieve the required reverb length by inserting small pieces of sponge rubber in between the mesh! Surprisingly it sounded quite natural and we used it for about a year. Finally common sense ruled and I won the day to purchase a stereo EMT 140 plate. A damping plate controlled by a servo motor, allowed adjustment of the reverb time from one second to approximately four seconds. It sounded fabulous – these plates are much sought after today – and we eventually had two. They both had personalities of their own – they sounded slightly different from each other, this was dependant on the tensioning of the plate. Our EMT (valve amp) plates were tensioned by Werner Wahl from Bauch – it was a precise operation to obtain the best reverb. Werner was very much respected by us in the London studio community and knew FWO Bauch from way back. We affectionately, some would say irreverently, called FWO "old man Bauch". He taught me much about servicing and valve changes in the Neumanns'.

Another major problem was the lack of any proper air conditioning; there were just Vent Axia air extraction fans. With the heat generated by the 100 plus valves in the equipment, and especially in the summer ambient temperature, the small control room was well above 30C degrees. I was determined we had to do something about it as soon as possible… As it turned out, that "soon as possible" was a long way off to 1970!

Outside of Denis's Record Supervision sessions there were new clients, plus old ones from IBC. Word got around how good the studio was: punchy, upfront and with the greatest clarity of recordings, especially for stringed instruments. Privately, in my mind, I still had doubts about the acoustics of the room for the way the work was developing with our "outside clients", and the changes I saw in the evolving music scene.

In January 1960, Denis employed a bookings person, Maureen Baker, who was session trumpeter Kenny Baker's ex wife so had knowledge of the industry. Prior to Maureen joining, Barbara Bray, Denis's secretary, was doing that job and Denis was a very busy producer, it was pressure. I required a person dedicated to that position. I thought Joe's departure would cause us a loss of business but fortunately it didn't.

In the 60s, the equipment was primitive compared to the wealth of technical wizardry available today. We now have a stack of outboard gear, 24/48 track recording and sometimes more, to DAWs (Digital Audio Work Stations) such as ProTools and Radar systems with software plug-ins and so forth. When Joe was sacked he left his "cooker" – quite useful! The HMV fan spring echo was dumped after I pulled it apart – no electronics, just a couple of telephone carbon-type earpieces taped on and used as send and receive transducers! Later, I eventually sold the cooker for £10.00 (£208.00 today) to Vic Keary. Vic resigned in March 1963, after a row with Stevens over his wages – didn't we all? Vic told me Joe's "cooker" was for sale recently on eBay for £7,000! I guess the Joe Meek connection made it valuable. Today, Vic's own company manufactures excellent valve outboard equipment.

To get an increase was like trying to get blood out of a stone. Occasionally I had run-ins with Stevens as well, always over money. One such memo dated 19th December '63 from Stevens read, *"I am very pleased to advise you that the directors have agreed to increase your weekly salary by £1 a week operative from the first pay day in January 1964...."* I had no say in the matter or any discussion with Preston or Stevens despite being a director! Engineer Dave Heelis (joined Lansdowne later in 1962) received the same memo as did other staff members. However, I decided to bide my time for the present. There were too many sessions to recall during my long career, but I can recall some of the more notable happenings and stories here, and in a later chapter.

Cliff Adams became a loyal client. I first started to work with him in 1960, when we pre-recorded some slots for *Chelsea at Nine*, broadcast from the Chelsea Palace Theatre by Granada TV. It was not an easy task with a twelve-input console in that hot control room, adjacent to a small separation room. Cliff could be quite intolerant with people, including me, and I was very wary of him. One had to get to know him. Cliff was a very prolific writer of jingles for TV adverts amongst his other work (incidentally, his singing group was the Stargazers). One such Jingle was the lonely man theme, *You are Never Alone with a Strand*, for Strand cigarettes, which at the time sold for three shillings and tuppence a pack – in old money. It became a minor hit in April 1960 peaking at #4 and in the charts for two weeks only. The BBC wouldn't play it and banned it when they realised it was advertising! Most of the musicians on that date were well known, Bert Ezzard (trumpet), Don Lusher (trombone), both players from the Ted Heath Band, Bert Weedon (guitar), Harry Pitch (harmonica), Ronnie Verrall (drums), Frank Clark (bass), and the strings led by very busy fixer and fiddle (violin) player Charlie Katz.

Another new client was advertising agency SH Benson. Howard Barnes from the agency was responsible for the SH Benson TV advertising, and he was more affectionately known as Boogie Barnes. He worked with Cliff and was very agreeable to have as a client. Cliff was in collaboration with Boogie on the writing of many jingles; I think I correctly recall some of the many the TV jingles I recorded with Cliff. *"Fry's Turkish Delight"*, *"All because the Lady Loves Milk*

Tray", *"For mash; get Smash!"* However, there were plenty other jingles and there was much work from that quarter which continued throughout the 60s. Siddle took over my work with Barnes and Cliff. Dave left in March 1963 to join SH Benson and Barnes, who were setting up a studio in Kingsway (Holborn area of London) to record jingles and voiceovers for advertising. Another major client was John and Joan Shakespeare. John was a trumpet player turned jingle writer, and his jingles were arranged by Ronnie Price, who was a most competent piano player/arranger. Ronnie and I became good friends when we travelled to Hilversum for the Shakespeares to record a jingle session in Philips Studios. After the date, we drank copious amounts of Amstel beer. Ronnie always mentioned that time when he came regularly for sessions at Lansdowne.

As an aside, I was approached in the summer of '67 to attend a meeting at the offices of Keith Prowse Music with Peter Philips, Bill Philips, Desmond Beatt, Joan Walker and Patrick Howgill. It was put to me that Humphrey's Holdings (a subsidiary of British Electric Traction (BET) and owner of Wembley Stadium) had land adjacent to Wembley Stadium and a new state of the art studio was to be built. Would I like to put it together? It was going to be called The De Lane Lea Music Centre, but became known as The Music Centre (much later CTS Studios). I politely declined and suggested that an engineer I had worked with, Dave Siddle, should be approached instead now working at SH Benson Kingsway Studios which had been purchased in the mid 60s by a Major Jacques De Lane Lea. I gave them his number. Although I recommended Dave, this offer gave me a chance to tell Stevens and Preston I was going to resign (a bluff) and set up a brand new studio in Wembley supported by two of our major clients – others clients would follow! I told them, "This new studio can be confirmed by those involved". The mucky stuff really hit the fan as panic ensued. Denis (a firm supporter in my camp) was furious with Stevens that he was about to let my talent walk. This prompted negotiation, not with Stevens but with the company lawyer – Ernie Whitaker – over a very good Italian lunch – during which I stood my ground with Ernie. A substantial pay increase ensued followed by a massive director's fee and a new five year legally binding contract. I knew the studio was making good money, lots of it, but was being skimmed off. Not one of the shareholders, myself included, were being paid a dividend. Stevens didn't make that mistake again. He obviously didn't understand that the talent of the engineers and the support staff made the business a success. Fast forward 20 years (1987) and I became the owner with Johnny Pearson of what eventually was renamed CTS Studios, renowned for its Oscar winning orchestral film work including many James Bond movies.

Word soon got around - since the success of Cliff Adams *"You're Never Alone With A Strand"* cigarette Ad - to other TV advertising agencies and at Lansdowne we found ourselves inundated with jingle work from many of the London agencies. J. Walter Thompson was one such big client whose Ad work was prolific. It should not be forgotten that commercial television then was in its infancy. One snag with that type of work was that it only required one hour to record a thirty-second (had to be precisely 29 seconds) or one-minute jingle (had to be precisely 59 seconds). Going over the hour triggered musician overtime! Musicians' Union rules. Ad agencies tended to book in the middle of the morning or middle of the afternoon, so as a revenue earner it was a non-starter. Bookings did their best to cram in a morning as many jingle sessions as possible, sometimes the first one (one hour) started at 8.00am and we carried on until early afternoon. Resetting the studio in between one-hour jingle sessions recording for a different agency was pretty much a "kick bollock and scramble". It meant we could get a single three-hour session in, say from 3.00pm to 6pm, and another from 8.00pm (we needed the time in between to set up) to 11pm.

On jingle dates a number of "executives" would often turn up to oversee proceedings and add their penny's worth of knowledge. On one occasion, the "executive" wearing an Army officer's heavy overcoat (which he never took off) and you could see on the lapels where he had removed the pips – an ex-Captain of gawd knows what! He kept offering his opinions but frankly he didn't know his arse from his elbow – hadn't got any "ears".

Early Techniques

I retained Joe's Langevin compressor (we had two) – I modified one and built a two-valve direct signal side-chain for it. The objective being that I could compress like a madman and allow some clean signal through. I used it on the DC5 hits in conjunction with the Marconi limiter – just one-quarter dB dynamic range on the finished master. I still have both these items. I also purchased from the states a Fairchild 670 limiter/compressor for £642.00. Much sought after today at an exceptionally high price.

I also built some direct injection boxes, having had wound custom-made transformers for impedance-matching from guitar and bass pick-up outputs, with a direct output to feed the instrument amp – common practice now. Impedance matching gives the advantage of converting the unbalanced feed from the bass guitar into a balanced signal for the console input stage. A transformer designed to handle very low audio frequencies requires very low *leakage inductance*. Air gaps between the windings are also needed to minimise the possibility of saturation. In those days, that meant a transformer with a lot of high permeability silicon steel inside, making it a rather large and heavy piece of outboard equipment, as well as expensive.

The lead from the bass guitar or electric guitar to the transformer was a low impedance-loss lead, whose inductance/resistance was low, mostly a coaxial lead – inductance and resistance will tend to minimise cable phase or time delays. This preserved the full frequency range of the instrument – i.e low phase shift and relatively little high-frequency losses.

I also insisted that the electric bass or guitar player use a high performance coaxial cable (from the guitar pick up to the DI or guitar amp input) with very low *transfer impedance* – a parameter left out of most cable specifications these days! The studio supplied the lead. A cable with low *transfer impedance* and low self-capacitance minimises any low – and high-frequency losses between the instrument and the console. In later years, I used an active self-powered triode valve DI bought from the US. A great-sounding DI box – no heavy transformers.

The long leads that players had with them were quite resistive and high capacitance. Some players said to me,

"You can't do that it will ruin my sound."

"I don't think so," I used to reply, "if you don't like I'll take it out and we'll revert to your old muffled sound. Have a listen on playback. I need clean punchy sounds that I can alter in the mix."

Some players took exception to what I wanted to do and used to say,

"I don't get this at so and so studio."

My retort was, "You probably don't get a good sound either and complain."

Silence! DI was relatively a new concept and the players were very suspicious the DI would be no good.

"I always change the sound on my amp"

"OK I'd say, that's not a problem we'll mic up the amp."

Satisfaction! Many times when the sound from DI was queried, although there was a mic on the amp, I never used it. On playback the musician was happy – it's all in the mind!

Some techie stuff for those interested: it is the combination of high cable capacitance and poor contact resistance in the connectors that makes low cost leads behave badly with poor shielding. The result of the cable capacitance and variable inductance at the connector contacts is that the impedance of the cable varies – sometimes even with the movement of the player – mechanically induced noise. This produces resonances that clash with the output impedance of the instrument. Some of you reading this may have heard it when at a concert – it produces a rattling sound in the amp when the player moves around.

Sometimes I took a feed from across the speaker terminals using an impedance matching transformer, the result was a noticeably twangy, wiry sound (not always used), which I mixed in either with direct input signal (not forgetting to reverse the phase on the console input) or with the miked-up sound. There were no guitar plug-in effects pedals then – although they're universal now. Many of the guitar players were creative and one effect was easy to create – the "fuzz effect"... The player simply overdrove the guitar amplifier into distortion. It sounded good! Also the amps had built in tremolo effects. There was one in 62 the Maestro Fuzz tone pedal used later by the Rolling Stones.

By this time I had decided not to use EMI tape as we received it in different manufactured batches. Audio tape, in simple terms, is a mixture of magnetic particles, resins and solvents coated on a thin plastic film – the result of what can be a very difficult manufacturing (coating) process in which consistency between batches could not be guaranteed. This was not acceptable to me as it made a huge difference when editing between takes that had been recorded on different batches. The effect on the top end could be noticeable. The tape could have frequency response variances, on line-up, of up to ±1dB, with the top end (15kHz) varying more like ±1.5dB. Variances in batch coating thicknesses and batch particle mix made the difference. Years later, when I was invited to go to coating plants such as those of 3M in the UK and Ampex in the USA to see the manufacturing process, did it come home to me how difficult audio recording tape was to manufacture while maintaining magnetic particle consistency, coating thickness and batch continuity. We changed tape supplier to AGFA, which had the advantage that we could receive it all the same batch. In addition the back of the tape had "AGFA" printed on it and the batch number – very useful during editing – and it would also accept a high recording level. This tape was quite abrasive on the machine heads. An added bonus to know which way was up if you lost an edit piece on the floor, it did happen!

Keith Prowse Music: (Not to be confused with Keith Prowse the ticket agency.)

Keith Prowse Music became a client of Lansdowne in 1961. They were recording library music for use in advertising and theme tunes. The first big band I recorded since the control room modifications was the Ted Heath Band. I admired the band, having worked on the sidelines once before with the stereo recording at IBC for the Americans – their record label was Decca. I liked Ted on first meeting him, although I was somewhat in awe. The sessions at Lansdowne for KPM were incredible – Johnny Keating the arranger and, I believe, the composer too. The modifications to the console and curing the phase problems with improved transient response gave so much attack and punch to the recordings that KPM were satisfied with the results and so was I! Ted could be a hard taskmaster to the band; they played well because they played regularly as a unit.

Ted returned again to Lansdowne to record a series of jingles with Johnny for the American tractor company Massey-Ferguson. I recall this because there was a *large* amount of 30- and 60-second jingles to record and no recording time constriction of one hour. I guess Ted could do this with his band; it appeared he had an agreement with the band to ignore the MU rules and pay the guys well! Would the guys object? I doubt it....

I also recorded a session of library music with composer Laurie Johnson, and one such piece called *Gala Performance* (engineered by me in 1962) became the signature tune for the TV show *This is Your Life*, with compere Eamonn Andrews. It was on air for a number of years. These first two recording sessions were the beginnings of many library sessions I did for KPM over 15 plus years with Robin Philips, including material recorded overseas (see Chapter 11).

As mentioned previously, gradual changes in the music industry were taking place marked by the emergence of a new genre of music – Rock'n'roll – and with it many new rock groups (bands). It took its time to arrive, since Bill Hayley and the Comets toured the UK in 1957.

Technical changes and control room move – Christmas and New Year 1962/3

As the studio became busier and busier we needed another engineer, and Dave Heelis joined us in March from Philips Studios, Marble Arch. Dave had worked for Denis Preston at IBC on a couple of albums in 1958 when Meek was excluded from working at IBC who edited the final works at Newman Street then Lansdowne. Dave was a very experienced balance engineer in many musical genres and, Philips was an excellent studio.

I wanted to get out of the small control room at studio floor level. I needed more studio space for the larger work we were recording, such as the Black and White Minstrel show – a tight fit in the studio. I wanted to move upstairs to what was supposedly (to be) the stereo control room but in fact was an empty room going to waste. Even before construction began on the new upstairs control room the TR90s were becoming maintenance heavy. They were replaced with AMPEX 351s - installed initially in the downstairs control room - and we sold the TR90s. The downtime, while the noisy building work was carried out, was ten days over the 1962 Christmas period into New Year 1963 with around two weeks for the tech install and testing. Our builder, Alf Williams, enlarged the upstairs control room window (we would from there look left sideways down onto the studio floor). It didn't make Lionel Stevens happy. He didn't understand the emerging technologies and questioned every penny spent, much to the annoyance of Preston and myself. I also needed to cover the old linoleum studio floor with carpet. Stevens, as usual, was tight with money and suggested we use a single layer of grey felt-like carpet. I had it laid and, oh disaster! The studio was a little deader but sounded awful with this thin carpet laid over the linoleum – a false economy. I insisted we rip up the linoleum and the awful thin carpet and use instead the underneath hair felt underlay on a different quality carpet of my choice to fit in with the studio decor. What a difference! The acoustic was not too dead and the sound of the room "felt right". A result that worked. Lionel typically questioned the necessity of various equipment. I recall one time Denis got very cross when asked once again,

"Why do we need this?"

"Because we bloody well do – you don't understand!" snapped Denis. The changes had to be done to progress the studio to cope with the increasing volume of work. One huge step that met with resistance, but was absolutely essential, was the installation over that December/New Year of the Ampex tape machines with the four-track Ampex 300, acquired after the 351s the two rack mounted 351U ¼" mono machines and 351-2U ¼ stereo machine and the modified EMI console with the newly built over-bridge and sub mixer for mixing rough mixes built by Peter Hitchcock. I had to verbally fight for its installation costs.

The move would also give us two isolation rooms at studio floor level. With the gradual changes in music genre, carpet was the answer for the studio floor. Obviously the studio would become "deader", but it was a chance I took. Brass players sometimes moaned because they were used to playing in live rooms, with mics not so close to the instruments as I placed them. When they heard the results the grumbling stopped. I used to say,

"I know the notes drop to the floor when they leave the bell of your instrument but just listen to your tight sound."

"But we can't hear anything from the other players," they said.

Simple solution – give them foldback! No headphones in those days, but small monitor speakers, similar to ones for stage foldback, before the advent of the in-ear monitoring now used by performers. The extra space created did indeed give us two isolation rooms, which I thought were required for the changing genre of music. To a certain extent, some of the long established studios were living in the past.

The Beatles at EMI with Geoff Emerick and the late George Martin (later Sir George Martin), and the DC5 at Lansdowne stopped all that. It was an era of the beginnings of large changes for pop music. Producers used to call me up and ask how I got those sounds – well, you see it wasn't a secret as such but it was commercially confidential. In simple terms, we used the right approach for the recording/s in hand according to whether it was orchestral or rock. As a musical fixer (contractor) David Katz, himself a violin player, once said to me, "It's all about horses for courses!" I didn't equalise very much. All the recorded sounds I made were created by knowing the shape of the sound pattern coming from the instrument/s, knowing the music score itself and where, for example, on a contrabass clarinet part, if the part was all written in the low end of the instrument I'd stick the mike in the top of the instrument to get that very big, fruity sound. You didn't stick the mike somewhere else where it didn't work – listen to an instrument and place your mike accordingly. So many people didn't do that. It has to be a musical approach towards engineering; it cannot be any other way.

Peter Hitchcock, a New Zealander affectionately called Hitch, had joined the company from EMI NZ in October 1962. He was an excellent experienced technical and recording engineer. When we moved the EMI console to the "new" control room we made a joint decision, and Peter would modify the console to build an over-bridge to accommodate recording, with a VU to monitor each track, to our newly purchased the four-track ½" AMPEX 300-4SSE recorder. We would build in foldback, which didn't exist before, because Joe had not originally specified it as he used to record "all up" so there was no need. I thought through the concept for monitor-switching the tracks. I drew out a block plan and Peter designed a system to switch the console outputs to any of the four track inputs and plug in direct to the other tracks if needed.

Also, a four-track sub-mixer would be located in the top right hand corner of the bridge to make rough mono mixes. We installed four Lockwood monitor speakers as seen in left, centre and right, and with one horizontally for track four. We removed the dual-concentric Tannoys and replaced them with Duplex coaxial loudspeaker 15" Altec Lansing 604Es, having a multicellular compression driver and passive crossovers – better sounding. The Lockwood cabinets worked well for them.

They were driven by very reliable valve (tube) H J Leak TL 25 amplifiers (25 watts) and it was sufficient power, as the Altecs were efficient devices. In that room, we were so close to the monitors, it was like wearing a huge pair of headphones! Large nearfields! The punch from them was amazing, especially the bass end. Air conditioning again didn't exist, only Vent Axia fans, and outside fresh air-changes served the room and the studio floor space. The heat generated by the all-valve (tube) equipment was considerable. We removed the two Lockwood's, for studio floor playback. A good reason for this was that when musicians heard a playback over on the studio floor they adjusted their playing, on the next take the balance changed. "I can't hear myself" was the usual remark which was not surprising as they were part of a well balanced band. Some musos were notoriously only interested in hearing themselves. On one particular date one brass player (the lead trumpet of the big band) complained. I invited him and other band members to the control room – the remarks were "it sounds great". Never again did we have problems – they were always welcome to the control room – a more controlled acoustic environment for monitoring (than the studio floor) and no more complaints!

Lionel and Denis agreed to the move – Lionel, being a bean counter, never wanted to spend money but Denis leant on him hard. This entailed a period of closure from 21st to 30th December '62 (a quiet time for the studio), to carry out the heavy noisy building work required to make the move, relocate the cable infrastructure, prepare the space, enlarge the control room window, and move the now-modified Meek console upstairs.

The move also involved installing new equipment racks from Imhoff, the 19" rack had hot air fan extraction, as well as the AMPEX tape machines – the four track, plus one stereo and two mono machines – installing the Leak monitor amplifiers, foldback amps and so on, all the while keeping the downtime to a minimum – no chance with all those modifications to do! It entailed long days for seven days a week. The acoustics of the room were OK. Denis told Lionel we'd make more money if we spent more money – and that was the clincher! And we certainly did make more! Denis told me that Lionel didn't understand our needs – so I had an ally. In Lionel's defence, he was a shrewd accountant.

Chapter 9
The Dave Clark Five – History and Techniques

Lansdowne was often booked by Bill Philips of KPM to audition new or up and coming groups. The raison d'être was that if the group was good enough the publisher would spend money on recording and get a release on one of the majors and publish the songs. This was the theory – usually the publisher took the lion's share! Most of the groups we auditioned were very poor performers and fell by the wayside. There were many of these groups whilst we didn't' realise it in the early 60s it was to become the decade of Rock 'N Roll when we irreverently called the bands – three guitars, drums, all sing; the songs based on only three or four chords, at least for rock bands. There was one exception to this pattern and it was the Dave Clark Five. I first met them when they came to the studio for demo work in 1962. It turned out that Lansdowne was the first proper studio the group had been in. With new inexperienced artistes not used to the professional studio environment, my approach was not to let them feel intimidated by the studio and all the professional gear. We recorded several demo songs in mono in a couple of hours. KPM did nothing with them but the demo masters belonged to KPM. I told the group that I liked what they were doing and spoke with Dave Clark:

"I have the studio, you have the group, and I am sure we can create a good sound. The material is there."

In particular, they were the one group that impressed me – my short-form name for them was the DC5. We agreed we could work together – Dave, with Mike Smith and the band. We would re-create their live sound, their audio signature. I had learnt early in my career how important the engineer was to artistes in painting the sound picture and giving direction about how to achieve the production technically, the end result recording and the final mix. Without sounding conceited, the engineer would help make a hit – hence Joe Meek being so much in demand. There are many examples of engineers creating sounds in cooperation with artists. One example I have in mind today is Bruce Swedien, an extraordinarily creative audio engineer, recording hit after hit for Michael Jackson with Quincy Jones – *Thriller* is a good example. Bruce is a very experienced audio engineer who recorded many genres of music including American big bands and jazz. I had the pleasure to meet him in New York at an Audio Engineering Society convention. Swedien started working under mentor Bill Putnam, another audio engineering great, at Universal Studios Chicago. Experience counts!

When I interviewed Dave for this book I asked him where it all began, and he explained, "*When we first started, we were writing songs and playing at American air bases in the UK, at dance clubs and on the Mecca ballroom circuit, which featured over 200 bands and catered for over a million people a week throughout the UK. These venues included the Mecca ballroom at the Tottenham Royal. We were recording demos of publishers' songs because that way they gave us free studio time to do our own songs. When we wanted to record demos of our songs, the only way we could do that in the studio was a little place the publishers used in Denmark Street, London, called Regent Sound.*

It was very sparse, they used egg boxes on the ceiling as sound-proofing and the recording was all made on one track. All the studio had was a one-track machine, but you went in and did the demo, and they would give you free studio time at the end of the session to record your own songs. We did some demos of the publishers' songs because they wanted a group sound and one of them got Gerry and the Pacemakers a Number 1 Hit, which was I Like It. *We did lots of demos and the big con on that was – nothing to do with Adrian or Lansdowne – was that we did the demos that they later released as proper records, but that's life. You learn by experience. It's part of growing but the great thing that came out of it was working at Lansdowne and meeting Adrian Kerridge."*

I learnt early in my career that teamwork is all-important when it comes to producing, and that the technique of production is not a process that happens independently from the artiste. There are some, in the UK and America, who think I did not work closely on production with the DC5. However, for the record, I did indeed do just that, working closely with Dave to create their very heavy upfront live sound – their audio signature – that was subsequently named "The Tottenham Sound" by the national press. Early records say on the label "produced by Adrian Clark". This was because the BBC in the UK refused to play records produced by the artist, so Dave came up with the idea of combining my name with his as the production credit, therefore enabling the DC5's records to be played on BBC radio. Once *Glad All Over* hit the number one spot, Dave was asked on live television news if "Adrian" was his brother. Dave said, "No, it's me!", but by then DC5 records were up and away, and topping the charts around the world.

Dave decided on the repertoire, and Mike Smith and some of the band also wrote songs. I took care of the recording and how we should go about it technically. Using only the Ampex four-track, musically planning ahead was essential – a team effort. I took time out to visit the Tottenham Royal to hear the group live. It was then, and always has been, my belief that with any band recording, and I don't only mean groups, not to record anything that cannot be reproduced live on stage, otherwise the music press hacks and the "knockers" will have a field day, especially when success looms large. And success loomed *very* large with the DC5!

The first released recording was *Mulberry Bush* on the EMI Columbia label. For me, it was getting to know the band and how to approach the recordings and, of course, the right song. *Mulberry Bush* was not it, and I was not happy with the recorded result – it lacked balls. It didn't reflect the big live sound they were producing at the Tottenham Royal , which was driving the crowds wild night after night. Back to the drawing board! It was important for me to establish the unique band sound that could be reproduced live, and this came not from me alone but by Dave's choice of material together with the indefatigable enthusiasm of Mike Smith and the group.

The DC5 were a clean-living group – no drug-takers in the band and they kept themselves fit by training in the gym. The great thing was that they were all friends from way back so there were no egos clashing.

After much thought about the required approach, we came to the conclusion the DC5 needed heavy, in-your-face sounds and a very much reduced dynamic range – I achieved just $\pm \frac{1}{4}$ dB on the finished cut. In those days, jukeboxes were everywhere in the coffee houses that many young people frequented. The machines had a fixed volume, so the louder the cut, the more the recorded sound stood out from the others.

I say this with great respect: the band was inexperienced when it first came to Lansdowne. It was a steep learning curve for them, and me, to achieve an instantly recognisable sound, but fortunately they were fast learners. In parallel with the studio work, the quality of their live performances was recognised in 1963, when the DC5 won the Mecca Gold Cup for Best Live band in the UK before they'd even had a hit record. I firmly believe that any band/artiste must have an instantly recognisable sound for continuity of success – their audio signature. Dave again: *"...our time at Lansdowne saw us working in the first proper studio we had ever been into. It was a fabulous atmosphere in the studio."*

The next step on the road to creating their signature sound was to get the band into the studio and record a series of songs. The Five always came to the studio well-rehearsed and later it was not unusual to record ten tracks in a twelve-hour day having established their recorded sound bar the mixing and maybe some vocal overdubs. In those early days, with limited outboard (audio processing) equipment and only four tracks to record on, the planning had to be precise. This was second nature to me, having worked only in mono before we had four tracks – unimaginable today with almost unlimited tracks to work with in a computer and, prior to that, 48-track or more analogue or digital recording locked in sync by timecode. Does having more tracks improve the sound? Yes, it certainly gives more flexibility but one can end up with a sterile sound without any feel – too many permutations and compromises for the producer. Great for studio income! Call me old fashioned if you will, but many of the old recordings of up to 50 years ago stand up very well today in their content and sound – such as the Beatles, Rolling Stones, Presley, Supremes, Crystals – *Da Doo Ron Ron* as well as the middle of the road stuff like Nat King Cole, Sinatra, and our own Petula Clark, Count Basie, Duke Ellington, Diana Ross and so on. Having the correct repertoire and recorded sound are, in my opinion, *two* of the most important ingredients.

After *Mulberry Bush*, with the band back in the studio, they changed their approach to the songs and I evolved my approach to the recording technique. The new recordings included *Do You Love Me*, an American song that the DC5 had been performing in their live act since playing at the American air force bases in the UK. We recorded this in August 1963 and it went well, and I knew when I mixed it that it would be a hit. This was not arrogance on my part, it simply sounded good and felt good, and had an aura of sound – a strong presence. Unfortunately, after the DC5 record was released, the song was covered three weeks later in the UK by Brian Pool and the Tremolos, who were following up a number one hit, so the record shops stocked their version of *Do You Love Me*. The DC5 version only reached the top twenty in the UK charts, but did stay there for six weeks, and went on to become a multi-million seller throughout the rest of the world!

With the DC5, I used to record as clean as I possibly could at 30ips on the Ampex 300 half-inch four-track recorder, using EQ and compression, where needed. For example, on the direct injected bass guitar I used a soft compression ratio of 2:1. I chose the all-valve Neumann mics very carefully, for their given sound output performance for different instruments.

A small example: U47 for Mike's voice, THREE mics only on the drums – U47 high overhead with the kit underneath a canopy, snare KM54 and kick U47, later changed to a dynamic-type Sennheiser MD21. I would take the front head off Dave's bass drum, and place a blanket inside the drum against a small portion of the rear head, held in place by a stage weight. The blanket's pressure could then be adjusted on the front inside skin to achieve the right amount of "firm" sound, and a hard or moderately hard beater on the kick pedal. This enabled me to place the dynamic mic inside the drum off-centre, using equalisation to achieve that thumping sound – no compression at this stage. Mike's Vox organ was direct injected. The instrument tracking differed for every take – there were no hard and fast rules.

Initially, I'd have an uncompressed rhythm track except for the bass guitar, with Mike's live vocal – for band performance guidance – on a separate track. I overdubbed Mike and the group vocals later. On the subject of the backing group overdubs, these would often be double or triple tracked with sometimes each overdub featuring subtle harmony changes suggested by me and worked out by Mike. It made for a huge sound. What about tracks I hear you say, there are not enough as you only have four! No problem! Planning ahead I could keep two tracks spare, and overdub the first performance, then overdub again to another track and so on, eventually wiping the not needed track using it instead for Mike's vocal. I would add any other instrument overdubs in the mix.

I'd say to Dave, "If we don't get it in three or four takes we'll move on and come back to it later."

Maybe there'd be a couple of false starts, but invariably we'd get a master within that time scale. That would have been a basic master with Mike's guide voice, but sometimes the live vocal would be used and the mic would pick up some of the ambience, which sounded pretty good. My approach to recording was simple – if you cannot make it in two or three takes forget it, as it is only likely to go downhill. But, when it happens right, as it did frequently, you feel good. If we didn't get it within those first few takes we'd go to the Mitre pub, have a beer and maybe lunch, walk back into the studio and bingo, done, in one retake! Ten titles from 10am to 10pm – that was a fantastic feeling. Nobody, but nobody, believes me when I say that today, but we achieved exactly that – the basic tracks, the overdubbing and the vocals. The reason for taking that approach is that the recording studio is the most expensive rehearsal studio in the world. Get it right, do it, come in and perform. And the other reason is that with Mike you needed to hear if he performed. You could not come in and do it to a click track or record the backing track only – it just didn't work. You had to have the performance.

Dave Clark on band mate Mike Smith: "*Mike was one of the most underrated great rock and roll singers. Mike didn't really realise how good he was... Looking back perhaps that was a good thing because there was no ego. We were all friends from way back. When we got the Tottenham Royal contract and were playing to six thousand people a night, which blew us away, I said to the boys, we will only go professional if we get two top-five records, we won't go professional unless we go out as the top of the bill act. If we are lucky enough and we do hit it off, we will stop while it is still being fun and that is what we did in 1970, going out on a million-selling record. Coming back to recording with*

Mike, the best bit of advice Adrian ever gave Mike was with all rock singers, especially with someone with a powerful rock and roll voice like Mike's, is that (mic) pop. And Adrian used to say, I don't want to use a pop shield."

Dave's comment is entirely fair. Mike was the only guy I knew who could sing half an inch away from the mic and not pop. The presence on the voice was phenomenal, and it made life very easy as he had a very controlled voice. He didn't sing through his throat and then down from his stomach, he sang from his stomach all the time and was relaxed – an exceptional sound.

I always used a valve U47 mic with the original Telefunken VF14M valve (later versions used a Nuvistor valve as the VF14M became obsolete and were no longer obtainable (It was said the use of this Nuvistor version changed the sound of the mic – I agree), and a small amount of compression, just enough to hold the voice but, as Dave pointed out, I never used a pop shield. Mike and I spent a lot of time together working on vocal technique, encouraging him to relax and project his voice. I put a lit candle between him and the mic, explaining that if he blew it out that's gonna pop, no blown out candle. I also said, "What I need is silent p's, b's and d's, but I still want you to enunciate." I'd also ride the voice, I'd have the lyrics written down and I'd ride the fader to get more intimacy or, if it was a bit louder, move it back. Mike helped by getting in closer when singing quietly which meant I could use the proximity effect of the mic capsule, a little more bass rise, better than eq. On loud passages, he would move back – he had a wonderful mic technique. By the time we were recording, he'd sing into the mic without a pop shield and it was perfect. On his voice I used a soft compression ratio of 2:1 and very little equalisation.

Every session with the DC5 was interesting; it was different. Sometimes for the weight of drum sound the drums were augmented by overdubbing, doubling up the snare sound with different equalisation or, even better, using a equalised gloved hand-clap to enlarge the sound, using some compression to hold the peaks. Another technique I used to enhance the sound to get more weightiness, was overdub a floor tom with a cloth over it, hit with a soft beater, and sometimes overdubbed twice to achieve that very up-front drum sound.

I was always very, very careful with level and processing. We used to record the instruments clean but equalised with compression where needed. Rick's bass guitar, which was a six-string bass, had just enough compression to hold the level. Sometimes I'd use chamber reverb on the drums and other instruments if required. I judged this according to the song, what my ear thought it should be or what felt right.

"Cooking" as Joe Meek used to call it, meaning, if you cook too much at the beginning it gets burnt and you can never undo it, but if you don't cook it enough you can always cook it a little more later. That was a very good lesson I learnt from Joe. If, on a particular song, you wanted, for example, Dennis' sax really raucous, you'd screw it up with the EQ, put a lot of midrange in it and maybe cut the bass with a simple filter.

The compressor I used was one that Joe Meek had manufactured by Langevin. I designed and built a two-valve (tube) modification called a "side chain", this enabled me to feed part of the mix uncompressed bypassing the compressor stage. I hear you techies asking why? A compressor with a side chain circuit allows you to use the output of one instrument (or track) to control (or trigger) the action of a compressor on a completely different instrument (or track). Simple answer: a controlled ratio of clean to compressed sound. Then maybe the Marconi limiter at 12:1 ratio, using it as a "brick wall". Now and again on the overall final mix to ¼" tape I would apply just a tad of final EQ and hey presto ±¼ dB dynamic range. The acid test on all my mixes was to check on what I called a grot box – a small cheap elliptical radio speaker – to see how the mix would sound over air.

We had the chamber and, for repeat echo, I used a mono Ampex, feeding it back to itself in the mix where required. On *Glad All Over* I fed the repeats into the mix in the gaps, actually physically pulling the repeats up with the fader. Timing was all. The DC5 sounds for instance, were simple; only five guys with overdubbing. For instance, adding Lenny on the *"Glad All Over"* guitar overdub fairly low in the mix, creating a tiny doubled-up guitar sound, which embellished the drums. It was those diminutive intricacies that made it special. When we finished the final mix of *Glad All Over*, I told the guys they had just made a record that was going to be a big hit. It excited me; it had an atmosphere and energy. I said, "That's your signature! That's what we wanted to achieve," and it went upwards from there. It was a massive hit! #2 all the way through Dec '63. The Beatles' *I Want To Hold your Hand* #1 in December '63 was their biggest-selling all-time single. The DC5 was #2 for four weeks, knocking the Beatles off the top position in the beginning of January 1964, and ended up selling 2.5M copies: nowadays you could get a number one on 10-20,000 copies. The DC5 were selling 180,000 copies a day – it shows the power of how big the record industry was then. It was top five in the United States in April 1964, and topped the charts around the world.

Remembering the teamwork on those sessions, I asked Dave to comment on the dynamics within the band, and on how the various personalities worked together. He explained: *"Looking back, especially compared to our contemporaries, I was very blessed because we've never had one legal letter between us to this day and that is over 50 years on. You cannot keep people in a band if they are not happy, and that's what pisses me and the boys off when people write negative things, many of whom weren't even born during the '60s. The boys could have left the band at any time; you cannot keep someone if they are not happy. We are great friends to this day. I believed that because it was a five-piece band you needed one leader; it's like a football team with a manager or the captain otherwise you have five different opinions and you don't get anywhere. Someone had to make the final decision but everyone had a say, whether it was Mike, Dennis, Rick or Lenny. What made it work musically were the little intricacies contributed by all of the boys. It wasn't just me; it was the five individual talents that made the DC5 work. It was all those combinations that made the sound. I just happened to be the guy at the front."*

The stomping on *Bits And Pieces* had more to do with fashion than technical wizardry. In those days they all wore Cuban-heeled boots and to effect the loud noise needed I suggested we stomped on the two wooden rostrums in the studio. I used a Neumann KM 54 microphone tight in to the boot heels, with all the guys stomping away, and used heavy compression with reverb, and over-dubbed it twice. The result was like a troop of soldiers all stamping their feet. It was great!

When I went to see the boys at the Tottenham Royal there was a sprung floor that bounced up and down with the stomping of 6,000 fans. It was tremendous – like being at Wembley football stadium, so exciting. The song peaked at #1 April 1964, and was in the charts for eleven weeks. In those days, you had four competing UK national singles charts, and *Bits and Pieces* was #1 in two of them, selling over a million records in the UK alone.

I asked Dave whether any other big selling bands had questioned him about the DC5 sound, and he said that the best compliment he received was when Freddie Mercury told him that the inspiration for Queen's *We Will Rock You* was taken from the stamping clapping stadium sound on *Bits and Pieces* Dave went on to recall, *"I told Freddie that in those days we stamped onto a board on one track, because we only had a four-track machine to work with, when Adrian was mixing down the master track. Freddie said they had so many tracks at their disposal, and overdubbed the stamping and clapping onto 20 tracks to get that big stadium sound, but our version of* Bits and Pieces *was their inspiration."*

I remember one Saturday we worked all day and recorded ten tracks, just basic tracks. Why spend two hours getting a drum sound? What's the point? I've seen engineers spend all day trying to get a bass drum sound. I always used to say if you can't get a recording going in the first 15 minutes and if you can't get a drum sound in 10, 15 minutes, forget it, even by today's standards. This applied to orchestral music recordings and orchestral "POP" where time could be at a premium. In later years (70s), when Hugh Padgham joined Lansdowne and then left to co-produce eventually Phil Collins, Genesis, Sting, The Police and others, he was frequently known to quote me about getting a drum sound.

During the 1980s, Dave received a call from Bruce Springsteen, asking how he could get the echo sound that is featured on the DC5 track *Anyway You Want It*. Dave added: *"It was the echo repeats in the song that made the track so exciting. Bruce, Steve Van Zandt and the E Streets Band's drummer Max Weinberg said they couldn't get the sound we had on the DC5 record. They had apparently tried to record it at three different times, over a six-year period. Bruce said in a recent filmed interview 'The DC5 made some of the greatest rock and roll records ever made – those were big powerful nasty sounding records, man. A much bigger sound than say the Stones or the Beatles… they were thrilling, inspiring, simply exciting – to this day they are still great productions,' which was very generous of him. I've always credited Adrian on all our record album releases (which included all the DC5 singles) for his technical wizardry. Phil Spector stated on The American National DC5 Television Special, 'The Dave Clark Five broke all language barriers with their phenomenally unique sound on record'. It's only years after that you take all this in. Berry Gordy Jr, the founder of Motown records and the Motown sound, stated that '…some of the early Motown records used Dave Clark licks (fills)', which was a great compliment! As an example, it's worth listening to the early Supremes records such as their first number one hits in mid 1964.*

Lansdowne House had a large stairwell reaching to the top of the building. I would run cables out for a mic and speaker and use the whole of the area as a large second echo chamber. The stone stairs helped to create a great sound; the only problem was if one of the building's tenants decided to walk down the stairs instead of using the lift/elevator in the middle of a take, we would have to record the track all over again!

We made very few edits and I preferred to edit the ¼" master after mixdown. Dave again: *"...even though the 2ⁿᵈ echo chamber was the stairwell to the building, when Adrian upgraded to a new state of the art echo chamber it was never as good a sound. Looking back, things were quite primitive, editing as an example was a risky thing because you had no option but to edit the finished master recording by making a Chinagraph pencil mark onto your edit points on the ¼» master tape, cut the edit section out with a razor blade and then splice the master tape back together. On playback, if the edit didn't quite work, you had to pick up all the ¼" master tape off the studio floor and piece it back together again. Today, with ProTools, one can edit sections without touching the actual master recording. Fortunately Adrian had good ears so nine times out of ten we got it right."*

As part of our discussion on the background to this book, I asked Dave about his production role and its wider context: *"As an independent producer, I was not restricted by the record company's rules, and they had to accept the master recordings I delivered to them. For example, the level of the recording should not have gone beyond a certain cut-off red mark on the playback monitor, otherwise the record company would have rejected it. The reason that the DC5 records were loud was because we ignored that rule and went way beyond that restriction as long as the track didn't distort. To me, that made the record exciting. You didn't have choices back then, hence the live sound approach we decided to take. With the DC5 records, it's the gaps we left on the track, so when you have a drum fill, guitar or sax lick,* it stood out. I believe it's the imperfections that make perfection and make that live sound come across on record. The danger today, with the luxury of so many choices, is that it risks becoming clinical. So many people around the world listen to The Beatles, Stones and lots of 60s UK and American records that now bridge generations, because they were exciting records, but one must not forget the songs. The songs, great songs, make great records and stay in the memory forever. As for the DC5, it was great to have Mike, Lenny and Dennis as my very talented song-writing partners."

I always believed that wherever possible we had to record a whole live performance, without overdubbing Mike, but only any small extras such as percussion or extra guitar, in this case Dennis' harmonica. There is nothing like a live performance. With a live performance the tempo can speed up a little because of the enthusiasm and energy in the playing. We recorded a song called *Over And Over*, in only two takes. The second take was the master for performance – it was a great take but a little too slow.

Dave asked me, "Is there any way we can speed it up?"

In those days, we didn't have Varispeed, but I said we could stick some Sellotape (or Scotch tape as they say in the US) around the capstan, and keep adding it on until we got it to the right speed although it would give us some wow and flutter.

"Basically, it's going to waver," someone, possibly yours truly pointed out.

We agreed to try it and got the right tempo but it did indeed waver – giving us that usually unwanted wow and flutter effect – particularly in the middle section on Dennis' harmonica solo. But… it actually sounded fantastic! Dave released it, and it went to #1 in America. Phil Spector called Dave and asked, *"How did you get that sound on the harmonica?"* Dave told me, *"I explained it was Adrian – it's a new technique he has. I didn't have the balls to tell him it was a five cent reel of Scotch Tape!"*

After that experience I decided we needed a better way to varispeed. I bought a high wattage power amplifier (for the techies, 12 x EL84 output valves working in push-pull class B), and connected the input to a signal generator set to 50Hz. I modified the capstan motor lead so it could be connected to the amplifier when needed. By varying the oscillator frequency by a few cycles – bingo! – we had varispeed. The amplifier generated sufficient output volts to drive the capstan motor. I used varispeed on other sessions at that time. This, of course, was before it became standard on analogue tape machines.

All the work we did was on the 4-track, all the way through – it was hard, but there was definition on all the instruments. It was using the ears and understanding the song musically, and understanding what we wanted to produce from it that made all the difference.

Planning ahead was critical because with only four tracks available, it was an essential part of "doing your homework behind the desk".

Later, the band was augmented live on the recording session with brass and strings on some album tracks – this was the first time we recorded and mixed in stereo. There are a number of the DC5 early cuts on You Tube purporting to be in stereo but they are not; we only recorded in mono in the early days. They have been manipulated and are pseudo stereo.

On one particular day, there were hordes of kids outside the studio – a small crowd at first, then it built up to hundreds, shouting for the Dave Clark Five. How they knew the band were in the studio could never be explained. They eventually BROKE into the studio, and ran amok – the DC5 ran for cover! The string players' music stands flew everywhere, and the musicians were not amused. Their leader was a fixer called Charlie Katz (you didn't upset Charlie), and they threatened to walk out and have the DC5 and Lansdowne put on the union's "blacklist" as it was called in those days – union hassles again. Stevens, our finance director, decided during all this ruckus to go home. Unfortunately, some of the schoolkids thought he was a member of the band and chased him up Lansdowne Road! Calm was eventually restored by the police, and the session resumed. Once the session had been completed, the DC5 had to be escorted into a police van and driven to safety, as by then a thousand or so fans were screaming outside the studio and it was live on TV news, headlined as "The Siege of Lansdowne Studios". Dave still has the black and white newsreel footage of that crazy day.

As with any success, especially in the music industry, the rumour mill goes into action, driven by those who abhor success. So it was with the DC5, especially as Dave was the first UK independent (pop) record producer and published all his own DC5 songs. No other artist had achieved this at that time, which didn't go down well with the old-school establishment. The rumour was that Dave Clark did not play drums on his records. Even today, I am asked that by some Americans who have an interest in the group: did Dave Clark play drums on his records? I noted that those who asked weren't even born then!

In the '60s, I had a call from the Daily Mirror, and the journalist asked,

"You record the Dave Clark Five?"

He was referring to *Glad All Over*.

"Yes I do," I replied.

"He doesn't play drums on his records," this guy went on to say.

"I have news for you, yes he does, and please don't ring me at home again!" was my reply.

Dave had a good publicist called Leslie Perrin. Les said he would invite six or so editors from the top major national newspapers and the music press to the studio, and he wanted us to re-record the songs. The band came into Lansdowne at nine or ten in the morning and worked with me right through until nine or ten at night. We had the major press there, and we re-recorded *Do You Love Me, Glad All*

Over and *Bits and Pieces* from start to finish, including all overdubs and mixing, on all four tracks.

Dave commented on that day's work: *"That all-day session killed the rumour completely, but now 40-50 years on, you get these people who weren't even born that read these rumours. All I can say is, if we weren't that good or we weren't that good musicians, we wouldn't have been playing to six thousand people a night, four nights a week at the Tottenham Royal, and awarded the Mecca Gold Cup for being the best live band in the UK, long before we made our first hit record. Playing live to that many people was great experience for the future."*

When the DC5 was at its height in the UK, I supervised the sound for Dave on many *live* TV shows (no pre-records then), including *Ready Steady Go*, and on 9th February 1964 when they topped the bill on *Sunday Night at the London Palladium* (the UK's equivalent of the Ed Sullivan Show), which established itself as the highlight television show of the week for Britain's viewing millions. The show immediately climbed to the top of the TV ratings, and was viewed live by over 25 million in the UK. At this stage the DC5 had not hit America and were still semi-professional, working in offices and factories with Dave also working as a film extra and stuntman. They subsequently appeared on two Royal Command Performances in the presence of Her Majesty The Queen and Prince Phillip, on Monday, November 8th 1965 and again on the 1966 Royal Command Performance.

On the live *Ready Steady Go!* shows, I sat in the sound gallery with the balance engineers, not a good experience! They objected to this "POP" engineer encroaching on their territory. They were polite, but I was not allowed to touch any of the faders on the board. Union rules, you see! I went about the job of describing what the band's sound was and how to achieve it. It was listened to politely, but it appeared they did not understand this pop band business and how to achieve the sound. I wasn't surprised at all, as the recording kit they had was pretty primitive – very little EQ, and one limiter

– not good at all. In fairness, they did the best they could with the equipment restraints they had. The director was creative but only cared about the picture shots. Very bizarre considering the show's slogan was "The Weekend Starts Here". It went out live at 7.00pm every Friday. I did several shows on which the DC5 appeared with the engineering guys gradually becoming friendlier as time wore on. On one TV show, there was a live band for other artistes conducted by Les Reed. I recall Sandie Shaw was on, well supposed to be, but just before her live performance of *Always Something There To Remind Me*, she disappeared to the loo (toilet) in her trademark bare feet, and didn't come back until her number was over. Les didn't blink an eye and continued to conduct the band top to bottom without Sandie. I imagined that viewers at home thought it was an instrumental version. It caused some disquiet in the gallery.

The Royal Command shows and *Sunday night at the London Palladium* went out live on TV and I was sat in a sound van in a back street, with equally old equipment. I was made welcome, but again not allowed to touch the faders. Fortunately, the engineering guys had some understanding, and for a live show with quick scene sound changes, they did a good job.

The relationship between Lansdowne and the DC5 was immensely productive, and the credits include:

- The film music for *Catch Us if You Can*, UK Release, and *Wild Weekend* US release were recorded at Lansdowne.
- From 1963 to 1970, all the DC5 records and albums were recorded at Lansdowne.
- 23 albums.
- Over 30 global hit singles.
- Over 100 million records sold.
- Over 100 Platinum, Gold and Silver discs awarded.

The DC5 were inducted into the American Rock and Roll Hall of Fame in March 2008 by Academy Award-winning actor Tom Hanks.

Chapter 10
Black & White Minstrel Show for BBC TV

Historical Background

The Black and White Minstrel Show ran on BBC 1 TV in a three-quarter-hour slot on Saturday evening prime time from 1958 to 1978 and at its peak had a regular viewing audience of 14 million plus. In the late '60s, the show received criticism that the "blacking up" of the male choir was offensive – even though it had been a long-standing tradition in the theatre. The girls in their colourful dresses remained white.

In 1967 the BBC received a petition by The Campaign For Racial Discrimination requesting the BBC drop the show from its schedules. The BBC finally did so in July 1978. It ran at the Victoria Palace then continued to tour theatres around the country for many years, as well as in Australia and New Zealand. The campaigners either were ignorant of the origins of presenting traditional American minstrel and country songs, as well as show and music hall numbers or chose to ignore historical fact. The arrangements were of 20th century songbook standards and those show tunes, plus traditional minstrel renditions such as *Oh Susanna*, and *Camptown Races*. The show generated wonderful guaranteed business for Lansdowne from 1961 to 1978, and we also recorded the choir and solo backing tracks for Victoria Palace Theatre, with the band playing live in the pit, and for the overseas shows. I personally never did consider it to be racist and obviously nor did 14 to 16 million regular viewers.

Recording the Show

In 1960, I am not sure of the exact date or month, Bookings had a call from the George Mitchell office to book in a recording session for the BBC TV production, *The Black and White Minstrel Show*, with the George Mitchell Singers (called "The Mitchell Minstrels"), and produced by a BBC producer George Inns. I was requested as engineer and recording the show entailed recording all the music, twelve male singers, of whom three were also soloists, including Dai Francis, who could sing in the style of Al Jolson, plus John Boulter (tenor) and Tony Mercer (bass baritone). The Minstrels also included eight girl singers – "The Television Toppers" – with Margaret Savage solo. There were guest artists including George Chisholm, Stan Stennett, Leslie Crowther and tap-dancer Benny Garcia, tap-dancing away on (one instrumental piece) a tap mat in the small separation room next to the control room. A regular guest singer with an impressive voice was Gerry Dorsey. In the mid 60s he assumed a stage name of Engelbert Humperdinck and had two massive hits in late 60s, *"Release Me"* and *"Last Waltz"* with many more to follow.

The band consisted of about 20 (it sometimes varied) musicians, all recorded in a four-hour morning session: not an easy task with only a 12-channel input board. I used one Neumann U47 on the men, who were raised on rostrums, on the girls, who were formed in a circle, a Neumann SM2 using both mic capsules in wide cardioid (the polar pattern changed electronically) with the top capsule rotated 180° from the bottom one. A Neumann U47 was used for the solo singers. Tall screens were employed to separate the singers from the band and the songs were recorded in separate segments.

After a lunch break, I had to edit the show making any musical performance edits required and some titles into extended segueing. This usually took two to three hours, depending on the complexity of the show. The show was recorded directly to mono, with no multitrack. I kept the tape levels high, with my console PPM frequently in the red and the Marconi limiter not working too hard to avoid nasty peaks; however, I used no tape compression. As you can imagine, there were many edits in the final master. A copy with no edits had to be made. The BBC didn't want any edit to fail on live transmission – a good precaution!

I didn't wish to lose any sonic quality when making an edit-free copy having taken so much care on the recording, with the mono machine correctly lined up. To maintain transient response, I played the master tape backwards, ensuring the playback machine was correctly lined up, as was the copying machine. The machines were connected directly together so as not to involve any extra signal paths. I wondered why George Mitchell wanted to record the show at Lansdowne. The BBC did not like to use outside recording facilities in those days. On one occasion, I received a call from BBC engineering, asking,

"Are you using a limiter"?

"Yes," was my reply.

"Which type?"

"Marconi (by Standard Telephone and Cables)."

"Well, its time constant is not compatible with our transmitter and your tape levels are too high. They go over PPM 7!"

"Oh, that should not affect the transmissions – just pull the playback level to PPM7!"

Pure sour grapes. We perceived it as BBC engineering politics coming into play. I heard no more about it! In researching for this book I found out their reasons for choosing Lansdowne. Pat Marshall, George Mitchell's widow, commented: *"It was definitely George who insisted that all recordings were done at Lansdowne because he said they were best at mixing and sound quality."*

Patrick Heigham, BBC TV Tech ops 1962-68, offered his views:

"The BBC assured George that all Sound Supervisors were capable of identical results, which wasn't good enough for him – he wanted the same guy on the desk for each session. Theoretically, we were all trained to the same standard. There were two Sound Supervisors at the time, who could have produced the results for George – Len Shorey and Hugh Barker."

Pat told me the BBC could not guarantee the same engineer all the time. It was wonderful business for Lansdowne, and I was tremendously proud when the show won the prestigious Golden Rose of Montreux Award in 1961 for Best TV Show including sound. The Golden Rose Award was the first international TV award and the show swept the board for best of everything, including sound, beating in America *The Perry Como Show* and *The Fred Astaire Show*. In its time the show was considered good, even classy, entertainment.

One disaster befell the studio. I arrived before 8am to set up for the show, a wired glass panel in the side of the studio door entrance was smashed; all the valve microphones were stolen overnight. What do we do now, a slight panic set in. I called up Keith Grant at Olympic explained the situation, he kindly sent over a collection of mics which got me out of trouble. We were insured and replaced them quickly. Stevens immeadiatly said we must install an alarm – horses and stable doors came to mind!

In 1962, George Mitchell announced the show was going on tour produced by Robert Luff, and I was to record the Minstrels at Lansdowne – backing tracks, choir and solos only. It would be recorded with a just rhythm section. The show content would vary; more than one recording had to be made. However, there must be no spill on the choir mics because the singers would be miming on stage to a pit orchestra. Thank goodness I could achieve that. George also requested: *"One other thing, we need to hear a count in."*

So it was decided one bar for nothing with drum rim shots. Crack, crack, crack, crack, then whip the fader down in a beat – easy. Then came the next surprise, the editing. First, the edited master tapes had to be copied at 7.5" per second onto 7" plastic spools – a no brainer. I copied backwards to retain transients.

"OK, that's fine," he said, and then came the crunch.

"How are you going to playback in the theatre?" – the Victoria Palace – "with stops and starts, as it's all edited in sequence!" I asked.

"We have a solution! We are playing back on modified Ferrograph recorders and all you have to do is to re-edit the copies and insert a piece of metal tape."

"Oh gawd, just like that!" I successfully edited in the metal tape, which involved a long, boring and laborious job.

"So what happens in the theatre?" I asked.

They had another solution, "The conductor has a foot pedal, the playback will stop when the metal tape passes an extra head. He will then start the machine for the next number when required."

Don't forget this was the '60s, with nothing like the technology we have today. After the show successfully started with full houses, I was asked to come to the theatre. Oh, what now..?

"We have a problem. We need to alter playback levels on some of the pieces and it is difficult to do and hear from stage left."

"OK," I said, "I have the solution. You need to install a small console at the back of the theatre and control the levels from that, and maybe have some reinforcement halfway up the theatre."

"We can't do that, it's too expensive and we'll lose seats."

"OK," I told them "You'll just have to manage as you are."

What a long way we have come in theatres today for the whole listening experience, across mixing console design, loudspeaker technology, sound-effect playout systems and overall show control. CADAC J-type consoles have been prevalent across theatres in London and elsewhere, and although relatively large-frame designs, necessitating the removal of some seats, theatres can increase the ticket price by a small amount to cover the loss of revenue. The trend is now towards smaller-footprint digital consoles.

Chapter 11
Into the 60's - from Acker Bilk to the Kray Twins

After the successful move from the old cramped and unbearably hot downstairs control room (no air conditioning!) to the new location upstairs we acquired more studio floor area to work with, giving us two isolation rooms. The old control room was now used as a vocal booth for solo voice or singers, or for a drum kit/percussion – ensuring clean separation when a string section was on the studio floor – or other quiet output instruments that required total separation. The original isolation room remained.

On completion of the new control room we reopened the studio without fanfare and a steady stream of work flowed in – we were ahead of the curve in terms of sound and attitude to the recording and our recording techniques. Artists trusted us to do the best for them and over the years through the doors at Lansdowne came a variety of jazz and pop artists including Moody Blues, the Animals, Procul Harem, Georgie Fame and Georgie Fame/Harry South Big Band, Lonnie Donegan, Craig Douglas, Emile Ford, Adam Faith, Joe Brown, Donald Peers, Norman Wisdom, Max Bygraves, Stanley Holloway, Vera Lynn, Tommy Steele, Peter Sellars, Elke Sommer, Springfields, Joe Brown to name a few. So many passed through our doors it would be impossible to list them all.

The studio won jazz awards from *Melody Maker* magazine for best recordings of the year. Awards in 1966 included first prize for *Jazz Suite – Under Milk Wood* by Stan Tracy and third prize for *Duskfire*;

both my recordings. Another of my recordings, an album by the Joe Harriott/John Mayer Double Quintet, the *Indo-Jazz Suite* placed high as did a recording by Georgie Fame and the Harry South Big Band.

Lansdowne Studios chart hits began from June 1959 with three created by Joe just before he walked out and subsequently sacked – his Lansdowne chart legacy. Two releases with Craig Douglas, *A Teenager in Love* which reached #13 in the UK charts and *Only Sixteen* which went to #1. Emile Ford's *What Do You Want To Make those Eyes At Me For*, also hit #1.

Later in the year I recorded Adam Faith's *What Do You Want (if you don't want money)* at Lansdowne #1 November 1959. The arranger was John Barry with Les Reed (piano) with rhythm section, chorus of singers' pizzicato strings and a small string section of six fiddles in the larger isolation room. How unusual, I thought. But in November 1959 it gave me my first Lansdowne chart hit! Those early hits brought more work to Lansdowne. Emile Ford returned to the studio in 1960 when I engineered a 10″ LP simply titled *Emile Ford and the Checkmates*, one of the songs, *Slow Boat to China*, went to #3. On that session the piano player was Alan Hawkshaw, a fabulous keyboard player and composer with whom I worked with on many sessions in years to come. Alan Hawkshaw: *"What is significant in my relationship with Lansdowne is that it was the first studio I ever recorded in, in 1960, when Adrian Kerridge was a lanky – and extremely enthusiastic – engineer there. I still find him very enthusiastic, though I don't know about 'lanky' anymore. I have worked there countless times. It's always in my opinion been a first class, high class, professional studio, with good engineers."*

When I last spoke to Alan he told me had managed to obtain an original copy of that LP. Emile went on at some length about a speaker sound system he had designed and would the studio be interested I had to be diplomatic and say not at present, I heard no more about it.

In 1960 producer Dick Rowe worked for Top Rank Records, a subsidiary of The Rank Organisation, as A&R (Artistes and Repertoire manager) after he left Decca records. His assistant was then the unknown Tony Hatch (later of Pye Studios and Petula Clark's *Downtown* fame, and many other hits). Dick brought much Rank Records recording work to Lansdowne. Tony also produced work at Lansdowne before Rank Records were taken over by EMI. I have one vivid memory of working with Dick. We were recording three titles (their names escape me) in a three-hour session. We did three takes on this particular song the first to be recorded that afternoon and the third take was, in my opinion, the master take. I said "That's the one Dick." He left the control room – pipe in hand – and walked onto the studio floor to address the musicians. He told them the take was good but "Can we have one more just like that?"

You can imagine the tittering that went on. One wag at the back of the studio piped up, "What's wrong with it then?" Musicians could be quite acerbic.

Dick held his ground.

"Well I'd like one more like that."

Nothing more to be said so the band played and the take was rubbish.

Dick just said,

"Thank you. Next!"

I turned to Dick and said,

"I'll mark the third take as the master."

Before he joined Pye Studios and while working for Top Rank Records, Tony Hatch became a fairly regular client of Lansdowne – a gentleman to work with, and a highly talented composer, songwriter and producer.

In 1961, Hayley Mills came to the studio and we recorded in mono *Let's Get Together* for the Disney film *The Parent Trap*. Her parents, John Mills (later Sir) and Mary Hayley Bell, came to watch the session – what absolutely charming people. At that time Mills had a string of films to his name and I greatly admired him – a superb actor. Those of you who might be familiar with the film will know the story is about identical twins separated at birth that discover each other at a summer camp; each one brought up by one of their biological parents. They make a plan to bring their errant parents together. In the film both twins sing *Let's Get Together*. Having no multi-track in 1961 I used the composite master technique for Hayley's second voice which meant we could record the "overdub" with multiple separate takes (mono). This was not an easy task for the young Hayley, as I don't think she had ever recorded this way before. It all came down to double tracking so timing had to be spot on – there were many second voice overdub takes but eventually it went well. I then had the task of choosing the accurate double tracking takes. In the end I made 32 scissor edits to get a perfect performance in the 2'.43" song. 1961 saw another Laurie Johnson theme – *Suku Suku* – recorded for TV spy series *Top Secret*. It reached #9 in the charts and over time was re-recorded in 16 different languages. Little did I realise then that I would spend over 40 years working with Laurie.

Laurie is a superb composer, whose scores always work though were often hard to record. He and Denis Preston worked together on many recording projects. Laurie recorded library music for KPM and his orchestras were usually larger than could be accommodated at Lansdowne. I recall one particular KPM recording in 1961 was at the Friends Meeting House Hall located on the Euston Road. The band consisted of a large brass/winds section and a huge amount of percussion. The hall had splendid acoustical properties. We used the PYE mobile recording unit. There was an issue about the tea break – not from us but from the tea ladies in the hall side room. Since the break wasn't when they were told it was going to be they were quite adamant if it wasn't soon there wouldn't be any tea as they were going home! That episode has always stuck in my mind. I sorted it with diplomacy to everyone's satisfaction.

Late in 1969 I recorded Synthesis at Watford Town Hall. It was a superb work by Laurie but with a combination of the London Jazz Orchestra, twenty hand-picked session musicians – some of whom were jazz artists in their own right – and the eighty-piece London Philharmonic Orchestra. It was a hugely complex project. Denis Preston and I discussed in some detail how were we going to record this combination of 100 musicians. I told him I would think it through. The answer was to use Bob Auger's mobile unit and *two* Neve "portable" (analogue then) 4-track output consoles and record to an eight-track Scully recorder – no Dolby noise reduction. Auger was to record the Philharmonic Orchestra using one Neve and I was to use the other for the Big Band. The outputs of the two consoles are combined to two monitors for listening in stereo. The 8-Track inputs each receives Track 1-4 from one console and Track 5-8 from the other. The recording was mixed back at Lansdowne on the studio's Scully eight-track.

Always, without exception, prior to recording I met with Laurie at his home to run through the scores describing each movement. Some of the movements were pretty black with the "dots". Not easy to record, I thought. The choice of the hall, then called the Watford Assembly Rooms, was based on an excellent acoustic in a large space. We discussed the orchestra layout and decided to place the Big Band on the stage and the Philharmonic in the main body of the hall. We used all valve (tube) microphones – some from Lansdowne and others from Auger. In the lunch break we retired to the local pub for "refreshment" with the some of the musicians. Bob always had his large gin and tonic – maybe two! I drank beer.

I visited Laurie with Denis's words ringing in my ears – "Do your homework behind the desk". When I interviewed Laurie for this book I reminded him about the importance of home work. He said:

"Preparation is everything. The great directors like Hitchcock would have prepared everything. One day he said to film composer Benny Herrmann on the first day of shooting as they sat in their chairs, 'Now Benny we come to the boring part of proceedings – the photography.' He had already made it all in his head. He would draw all the set-ups on the left hand side of the script like a storyboard so he knew exactly what he wanted. Bob Berkshill, the cameraman, would look through it and Hitchcock would leave the set up to him – he would just follow that blueprint. That's all preparation".

How true: Once the prep was done Denis and Laurie would leave it up to me. I had the main parts of the scores in my head – I knew what was coming. On large orchestral jobs I would ask for a copy of the score in advance, more homework, freeing Bob and me to concentrate on the technical aspects on the day of recording. I asked Laurie Johnson about this: *"Being a composer or a playwright you have this middle layer, which is the interpretive layer, so the creator is the composer or writer and that is it. The interpretive art comes from the musicians, actors, directors. Which you have to hear through your head all the time so when I started doing music and recording that process seemed obvious to me. You have to remember I used to get to a recording session about an hour before anybody else. It was like casing the joint, to make sure everything was in place."*

Laurie on Lansdowne: *"Adrian Kerridge, the engineer at Lansdowne Studios, is brilliant at what he does. If you're going to do background music using bands, orchestras and strings and so on, there's no finer engineer or studio to use than Adrian at Lansdowne. And that's how that started."* When Laurie worked as a composer for KPM it was Laurie who recommended me to Robin to record the KPM library music.

Another huge recording work I did for Laurie at Watford was an album called The Conquistadors recorded 16-track non-Dolby and mixed at Lansdowne. It is worth mentioning here the work is scored for ten trumpets, ten French horns, ten trombones, two tubas, eight percussion instruments and an organ. There was a thunder drum on one part of the score and it had to be played with very quietly "ppp" (means "pianississimo" and played "very very softly"). I miked it up with a KM54 close to the skin on the outer edge of the drum to achieve the desired effect of a distant murmur of thunder. How well it worked in the mix – no synthesisers then! How imaginative of Laurie's thinking in the scoring. A composer who knew and thought about recording techniques of the time.

The organ at Watford was a Compton theatre organ originally from the Gaumont Palace Theatre Chelsea, a three manual pipe organ with 16 ranks and many stops, some of which provided percussion instrument sounds. It could therefore be used as a straight organ or a theatre type with effects.

The score required in some parts – there were many bars of rests to count – low sub bass pedal sounds. During the recording we were in the box following the score and renowned musician Harold Smart was sat at the organ. We reached to the appropriate place in the score for the organ and there was silence... no organ. I thought Harold must have lost count so my assistant went to check and found Harold asleep at the organ despite all the brass noise around him! We made an edit and said nothing to Laurie.

The mixing was carried out at Lansdowne and in the case of the Conquistadors we recorded the narration by Sir Bernard Miles separately on ¼" tape at Lansdowne and then edited it into the 16-track post mixing. Throughout the '60s Denis Preston's work continued apace – Record Supervision commanded a fair amount of studio time. He also recorded demos of artists with a view to recording them if they proved good enough. One such demo was that of Gerry Dorsey (*Black & White Minstrel Show* regular guest artist) recorded by Vic Keary. Vic takes up the story: "*My first was recording Gerry Dorsey, a piano and voice demo. We got on quite well, both having an Indian upbringing so this may have made him pretty relaxed."*

Afterwards, Denis asked Vic to take a copy of the demo to Esmeralda's Barn in Kensington – the nightclub owned by the notorious London gangsters Ronnie and Reggie Kray. Vic recalls: *"When I got there, early evening, I was greeted royally by a very attractive hostess and offered a glass of Champagne to drink whilst Ronnie and Reggie listened to the tape. They were very pleased and instructed the hostess to give me a packet of sandwiches from the fridge to take back. The package of course contained cash.* What Denis was doing dealing with the Kray Twins never came to light. Denis kept such matters to himself. It is a well known fact the Kray's mixed with 60s artists and entertainers. One can only speculate about the cash! Gerry Dorsey became known as Englebert Humperdinck. (Chapter 10, *Black & White Minstrel Show).*

Acker Bilk and the Paramount Jazz Band was one of Preston's jazz bands artists who were successful in their own right. I recorded the bands first hit late 1959, *Summer Set*, which peaked at #5 in the charts 16th January 1960.

In 1961, Denis changed tack with Bilk – Acker was a West Country name for "friend" or "mate". One of Denis' many talents was to coalesce unlikely scenarios with artists. He wanted Acker to record with strings and rhythm. He chose Leon Young, the arranger, orchestrator, pianist and organist. Leon was a staff arranger at Francis Day & Hunter, music publishers, in Charing Cross Road. The generic name for a series of album recordings was Acker Bilk and The Leon Young String Chorale. However, we continued to record the Paramount Jazz Band. In 1961 Denis told me his recording intentions with Acker and suggested we record the strings with a lush "open" sound. To do this the answer was to use a carefully placed Neumann stereo SM2 over the whole string section and a back-up mic for the cellos if required with a mic for the harp (KM54) – and my established mic technique for recording for the rhythm section – (only three mics on the drums one U47 one metre overhead the kit one KM54 near the hi-hat and one dynamic on the kick). Previous to the date it was known Acker would not be available. Denis arranged for Sandy Brown as a stand in for Acker, I isolated Sandy in the separation room (ex-small control room, now isolation room).

Recorded on 4-track, strings and harp in stereo, rhythm section on a track and Sandy on the remaining track. Later we over dubbed Acker on all tracks in no more than a couple of takes.

The tune was written by Acker for his daughter Jenny and that was the title on the musicians' score parts for Acker's album *Sentimental Journey*, the title was later changed to *Stranger on the Shore* for a BBC television series. Released in October 1961 (Columbia records DB 4750) it peaked at #2 in the UK and stayed it the charts for 58 weeks. It was the biggest selling UK single in 1962. In America in May 1962 it was the first British recording to reach number one on the US Billboard Hot 100. Leon is no longer with us. His son Malcolm on the recording: *Adrian Kerridge was "the engineer with an ear". "The contribution of the engineer cannot be readily dismissed. Over and over again we hear of the interpretation of such-and-such a conductor but the engineer can control the overall balance and bring out the pertinent instrumental emphasis far more than can the conductor. The 'interpretation' lies within his hands."*

1962 saw Nina and Frederik recording to our newly acquired Ampex ½" four-track that was when I met composer arranger Syd Dale. Again, little did I realise we would be working together for the next 30 years. Frederik was heavily into pot in the evenings and frequently the session – voice overdubs – had to be abandoned. I recall one particular evening Frederik was leaning against the studio floor wall and he slowly slid to the floor – totally out of it. Nina just put up with it, at least in public. When working all day she insisted we go back to their Knightsbridge apartment and make us lunch.

Having worked on a couple of Nina and Frederik projects Syd Dale contacted me to see if I would come to his house to meet a guy called Chris Blackwell. Syd told me Chris was the former *aide de camp* to the Governor of Jamaica, Sir Hugh Foot. Blackwell was a producer who wanted to restart his own record company in the UK, Island Records, which he founded in Jamaica in the late '50s recording Jamaican popular music. He had a cassette tape he wanted me to listen to which I did several times. He told me he wasn't happy with the mix he'd done at Olympic Studios in Barnes, recorded and mixed by Keith Grant, (that surprised me because Keith was an excellent engineer) or the vocal performance. He asked what could I do. He wanted to replace the vocal and remix. OK, I said, no problem. The master multi-track was four-track – compatible with our Ampex 300. Chris booked a half session in early 1964 from11:00am to 1:00pm.

He brought in the master four-track which had only one track spare for the voice and I erased the old vocal performance track 4. The singer arrived, and she made a couple of performances dropping in vocal corrections where needed – she sang well and the vocal was mastered in a couple of takes.

Then Chris told me he had a harmonica player coming in to overdub a solo in the middle of the song where there was no vocal. I told him it would be a tight drop-in and out so the harmonica solo had to be bar perfect in length every take. The player, a scruffy looking guy, arrived listened to what he had to do, one rehearsal, and in two takes that was it. My drop-in and out were perfect. I set about mixing the track using equalisation and compression. Job done: from start to finish in just under the two hours.

The artist was Millie Small and the name of the song was *My boy Lollipop* – Philips Fontana label – recorded in Jamaican "Ska" style entered the charts in April 1964 and peaked at #2. And the harmonica solo? That was Rod Stewart. I then recorded *Sweet William* with Millie, which entered the charts in June 1964 peaked at #30 and was in the charts for nine weeks. This was the early beginnings of Island Records in England.

Chris brought much work to the studio with his business colleague Graeme Goodhall. Three names that come to mind are Jackie Edwards, Jimmy Cliff and the Spencer Davis Group for whom I recorded *Keep on Running* (others have claimed they recorded it) amongst other songs. *Keep on Running* entered the charts Christmas week December 1965 peaked to #1 for fourteen weeks in UK and to #2 in the states, written by Jackie Edwards, then *Somebody Help Me* entered the charts in April 1966 peaked to #1 for ten weeks. This track quite famous for its fuzz pedal guitar sound. The pedal was a Fuzz-Tone FZ-1 also used by Keith Richards on *Satisfaction*.

Jackie Edwards was principally a songwriter from Jamaica; however, he was also artist in his own right for Island Records and I recorded many titles with him including *Come On Home*. I also recorded reggae for Chris. In 1967 Muff Winwood (brother of Steve Winwood) joined Island Records with Chris Blackwell as an A&R man. He produced for Island and booked Lansdowne for many of Island's productions. In early the 1970s Blackwell opened his own studios, Island Studios in Basing Street, near the Portobello Road not far from Lansdowne.

Balance engineer Dave Heelis joined Lansdowne in March 1963 coming from Philips Records in Stanhope Place, Marble Arch. Dave previously worked on a couple of albums for Preston at IBC: *Sonny, Brownie and Chris* (Sonny Terry, Browne McGhee, Chris Barber) on Nixa Jazz Today Series recorded at IBC 2nd May 1958. The second album was recorded on 7th and 8th May 1958 for Preston and was entitled *Sonny Terry and Brownie McGhee*. In London, Joe reduced to the editing the recordings in the new part constructed Lansdowne control room. Dave was an excellent balance engineer and a real gent – nothing was too much trouble to do to achieve an end result. During the period after Joe left IBC he was not allowed by Stagg to go to the studios to record hence Heelis and Eric Tomlinson did the work until Lansdowne became operational.

In 1964, a group called the Barron Knights was signed by Denis, Their recordings were engineered by Dave Heelis, ex Philips engineer, and between 1964 to 1968 they had six chart hits. I think they appealed to Denis's dry sense of humour because they parodied the groups of the day including, as one example, Dave Clark Five whose *Bits and Pieces* was parodied as *Boots and Blisters*. They also parodied Freddie and the Dreamers and the Bachelors: and they got around copyright restrictions. The first of their six hits was *Call up the Groups!* The single peaked at #3 in the charts 18 July 64.

Producer Gerry Bron, whose clients included Uriah Heep and Motorhead, became a huge client of Lansdowne. I knew him well and sometimes was invited to his home for dinner with his wife Lilian. He brought to Lansdowne John Heisman's Colosseum with Barbara Thompson. I had recorded Jon before when he played drums on a Mike Taylor trio session LP *Pendulum* (a Preston session) I recorded in '65. I am sorry to say the session didn't start off too well. I think he had forgotten we had worked before with Mike Taylor. John's drum kit was large, which was not a problem except he had two bass drums (kick drums) and started to tell me how to mike them up as well as the rest of the kit. I politely listened. He said he wanted two U47s on the kicks, about three feet from the front heads. "Hmmm..." I thought "You're the artist so I will give you what you want..." but I knew from experience this wouldn't work. He also used soft beaters on the pedals – even worse! We further had a discussion about the mics to be used on the kit. He wasn't sure so I told him leave it to me and if he wasn't happy we could experiment and change the mics. My woes were further compounded when the bass player, Tony Reeves, came to the control room and asked me if he could equalise his electric bass as he knew the sound he wanted.

He promptly started to twiddle the knobs, quoting the amount of equalisation lift and the frequencies used. He was adamant he didn't want to listen to the recorded bass first, which I found bizarre. They had not played in this studio before and all studios and equipment are different; so is the result. Working under those constraints I got a sound together as much as I could. My whole approach to recording was to get a punchy upfront rhythm sound and I knew that if that worked the rest would sit nicely. Not in this instance with the flabby kicks and bass sound. Playback time! John was horrified and most probably thought this engineer guy was useless.

"That's not good," he said.

To which I replied,

"I agree and nor is the bass."

I could foresee the whole session going down the pan very quickly. In fairness to Gerry he told John to trust me – thanks Gerry – and John asked what I wanted to do.

I wanted to take the front heads off the kicks. That suggestion met with an outright refusal resulting in a long discussion.

"John trust me I know what I am doing," I said and, reluctantly, the front heads came off.

"I also am going to change the kick mics to dynamics and stuff a cushion on the back head inside the drums not over damping the back head; we'll listen and adjust for a solid sound using harder beaters on the pedals."

My final suggestion, "Let me sort out the electric bass" was deemed another problem.

I persuaded Tony Reeves to let me DI the bass, thereby getting rid of the long stage lead (These leads were cheap – made with high capacitance coaxial cable and connectors that had poor quality contact material) and instead used my low-impedance low-capacitance lead and my special DI transformer. To keep him happy, I miked up the bass amp as well to capture pedal effects when he used them as on

Mandarin. This also gives the advantage of converting the unbalanced feed from the bass guitar into a balanced signal for the console input stage, also applies to any direct inject instrument.

Tony played without a plectrum, giving a good funky sound and a whole new bass sound with the DI. There is a long bass solo on *Mandarin* where you can even hear the string movements against the fingerboard

Listen to *Walking in the Park* and the lousy whumpy muffled bass, in fact the whole track, in my opinion, is not technically good: wasn't me guv! I got a drum sound together in about ten minutes by the time I had made various EQ adjustments, then I sorted the bass mess out with different EQ: and we made a take – all was well.

I learnt this was their first album, *Those Who Are about To Die We Salute You*. An apt name, I thought, considering the happenings on that 1st session. I only recorded a few tracks on that album. The various follow-on sessions were a breeze, as were the mixes. Hiseman on Lansdowne: *"My associations with this studio go back to my beginnings in the music industry. I made two LPs with pianist Mike Taylor there in 1963/4, and subsequently went back there for most of the original Colosseum recordings (with Gerry Bron producing and Adrian Kerridge engineering).* "I'm still often asked to sign copies of those recordings, particularly the famous* Valentyne Suite *and a couple of months ago, intrigued to see how it sounded, I played the record for the first time in eight years. The cohesion and musical quality of the sound quite took my breath away, despite all the technical advancement of the years between. All I can say is, that Kerridge bloke certainly knows a thing or two"*

Gerry had so much work and because I was also managing the studio and staff it was impossible for me to do it all. Later on, with Gerry's approval, I handed the work over to another of our guys, Peter Gallen, who went on to record Uriah Heep, Osibisa and other Bron artistes. Peter was a capable engineer and when he left Lansdowne he joined Bron at Bron's Roundhouse Studios – the record label there was Bronze Records.

In 1963 I worked with Gene Pitney on *24 hours from Tulsa*, with Gerry Bron as the producer, which peaked at #5 and stayed in the charts for 19 weeks. It peaked at #17 in the US. I used plenty of reverb using our echo chamber! From my past memory, it wasn't a problem to record. The playing in *Where the Sun Don't* was a fad on practically every session in the mid 60s: the use of bass guitar and upright bass playing in unison: Bert Kaempfert got it right on his Albums. Only a few players knew how that worked for sound, otherwise it was click and boom, whumpy sound and the usually loud bass guitar sound rolling around the studio. I frequently requested the bass guitar (politely), do you think you could turn it down to a deafening roar! Some guitar players were just as bad – no sense of balance within the band.

I remember another session arranged by Kenny Salmon when he was on Hammond organ. The artist was well known as she had appeared in some *Carry On* films. The band arrangement was very busy with brass, rhythm section and organ. The arrangement was very much in your face. Looking at the arrangement it was black with the dots (notes) – known by us colloquially as "fly shit" – not many gaps for the voice.

When we overdubbed the voice later the artiste's manager complained the voice was swamped by the band and demanded more vocals

"More vocal is not a problem but the backing will have to be pulled back", I replied

"No I need both. I need more voice. Why can't you do this, what is wrong?"

Now came the crunch time,

"Well she doesn't project enough to get the voice over the backing. I can pull the voice more forward at the expense of the backing – which you don't want."

My point was made politely. I wasn't rude and we left it at that; I did the best I could considering the conflicting instructions. I didn't see the lady again for years and when we did eventually meet she greeted me ...like an old friend! No hard feelings there and lessons learned by both.

In later years when a producer asked about the voice levels and asking whether he could control the voice on the overdub the answer was always yes. The engineers came across this frequently so we adopted a technique. We gave them a fader to push, known as the DFA (Does F**k All), usually down the far end of the console, which was handy for the producer! Oh happiness, a producer justifying his job. It worked wonders. It enabled us to get on with the job following the lyrics and riding the voice with the producer too engrossed in pushing his fader. In the early days most producers I worked with were very good at their jobs and confident about letting me get on with it. But others were hopeless. How they got the job was difficult to comprehend; some didn't read music or heard one thing in their head and described another. Some were just full of BS. We saw through that but all the time we kept our "bedside manner".

In those very early days there were no dedicated assistants or techs. It was down to me or Peter Hitchcock or Vic Keary to set up the session be the tape op, record, do the maintenance and machine line-up.

Carlton Facilities came to Lansdowne from IBC thanks to Desmond Beatt and producer Joan Walker. Recording background music was good, consistent bread and butter work on a regular basis, planned by them and booked weeks in advance. I recorded many well-known names of the time including Frank Chacksfield, Sid Philips, Tony Osborne (trumpet player, pianist, composer, arranger, conductor), Malcolm Lockyer (British film composer and conductor), Ronnie Aldrich (Decca artist, composer, conductor, jazz pianist), Ted Heath (Big Band leader), Syd Dale (composer/arranger former piano player with the Squadronaires), Harold Geller (composer/ conductor) and Eric Winston. Desmond Beatt wished to include Lansdowne in its brochure sent out to all their clients, good free publicity. Carlton Facilities became part of Associated-Rediffusion, which was originally a partnership between British Electric Traction (BET) and its subsidiary Broadcast Relay Services Ltd and owner of Wembley Stadium Ltd and Humphrey's Holdings Ltd. Carlton Facilities traded as Reditune.

Joan Walker produced all the sessions (a first rate producer with good musical ears, and an excellent knowledge of music – it made a change!) with Rediffusion artists and also the Frank Chacksfield Orchestra. Chacksfield was a Decca artist, whose albums were released as "concept" LPs. Young wrote the arrangements for the Chacksfield Orchestra, usually made up of 40 musicians, and conducted most of those sessions at Decca. Typically the Lansdowne sessions saw two sessions a day with twelve titles recorded over those sessions providing the total music recorded didn't exceed twenty minutes in any one session – Musicians' Union rules you see! All session players on the London scene were more than capable of that, as they were excellent musicians. A few unionised players quietly timed the amount of work recorded and questioned it with the conductor. Mostly however the players knew which side their bread was buttered and turned a blind eye. I knew of no musical director who would deliberately take advantage except one. Frank was an arranger; personally I never thought Frank was a competent musician who appeared to have talent. We insiders knew that Frank did little arranging – that was Young – and it was most convenient for Frank to let people think the work was by his own hand. I recall one date there was a copying mistake on a part. I heard it and knew where it was and so did the guys but nobody would tell Frank.

Frank called a tea break and spent the fifteen-minute break at the piano trying to sort it out. He found it eventually in the trumpet part. The player's didn't respect him that much and as a result were not cooperative. One bugbear for me was the hour and a half lunch break at Frank's insistence. Frank always paid for lunch at a restaurant over a pub in Notting Hill Gate. He always drank white wine chilled in an ice bucket brought to the table. When the waiter came to open it Frank felt the neck of the bottle and told the waiter the neck wasn't cold "put it back with the neck immersed in the ice." The wine was eventually opened only when the whole bottle was satisfactorily cold. He was hugely successful with his orchestra on Decca label. One other thing amused me he would talk about climbing into a bath and jumping into a shirt, why this came up occasionally I don't know but it did.

Chacksfield's recording of Charlie Chaplin's theme from *Limelight* was arranged by Leon Young – as was *Ebb Tide*. Both the recording arrangements by Young gained Chacksfield two gold discs, one in the US and one in the UK. Recording the Sid Phillips band was interesting. An English jazz clarinettist, bandleader, and arranger whose record releases were on His Master's Voice label (EMI).

Phillips was a disciplinarian with his players and worked them hard. Prior to embarking on the series of sessions, he liked to meet with me, view the studio, discuss the recording/s and his band's sound, which was fair enough. When we met he told me he only used one mic for the band front line and they were to sit in a circle with the mic in the centre. I think he worked like that because he was an EMI artiste recording at Abbey Road and that was the way they did things. I wanted to mic up the bass, piano and drums, which he accepted, with a separate mic for him. I wondered what I was in for! For the front line I used a U47 in omnidirectional a KM54 (tube) on the bass and two U67s on the piano and a couple of mics on the drums. It actually sounded good. Thank goodness our studio's reverb time was not large otherwise it would have sounded dreadful. On another session Sid came with his 11-year-old son and asked if he could sit in the control room for the session. His son wanted to see a session in progress as he was learning to play drums. His name: Simon Philips, now a prolific session drummer who has played for The Who, Toto, Mick Jagger, Mike Oldfield and 10cc among others. The Ted Heath Band had some marvellous players. Band members included trumpets; Bobby Pratt, Bert Ezzard, Eddie Blair and Duncan Campbell; saxophones, Les Gilbert, Ronnie Chamberlain, Bob Efford, Henry McKenzie and Ken Kiddier; Trombonists, Don Lusher, Wally Smith, Johnny Edwards and Ken Goldie; drums, Ronnie Verrell; double bass, Johnny Hawksworth; Piano/Percussion, Derek Warne. In the years to come I worked with many of these splendid musicians on session work. It was always a pleasure to work with the band. When conducting Ted fronted the band sideways and to the right – his back never to the audience. Ted always came into the box to listen to playbacks and, being a trombone player, he always wanted plenty of trombones but this could make the mix not in balance and trombone heavy. The trombones were usually placed in the mix left to centre, the trumpets right to centre with a good brass spread. He always said "More trombones! More trombones!" even though in the mix the balance was correct. I found the trick was to raise the audio level of the left monitor until Ted was happy. This puzzled me until I realised all the years of standing in front of this marvellous swing band had left him hard of hearing in his left ear – all that brass in his left ear from years of conducting sideways!

We got on very well together; he called me up one day in the middle sixties to say his radiogram had broken down and could I fix it? I went to his house and met him and his charming wife Moira. The radiogram was a huge floor-standing beast. I found a faulty valve (tube), replaced it and extracted heaps of dust that had accumulated over the years. Everyone was happy!

All Carlton sessions were recorded direct to stereo unless the recording required multi-track, which was not often. Ronnie Aldrich, former leader of the RAF Squadronaires band, and his two pianos were an exception to the rule and recorded in multi track. He used arrangements written for the Decca Phase Four series; two pianos, piano left and piano right. Aldrich would conduct the band in the morning then in the afternoon we would overdub the two piano parts. In the one-hour lunch break we would go to *The Castle* on Holland Park Avenue for a sandwich. Joan Walker and I would have a small beer but Ronnie would usually have two or three large gin and tonics. It never impaired his performance in the afternoon. Now there was an excellent musician, composer and arranger!

There were times when we had to turn work away or the client would wait until I or another engineer was free. The studio philosophy was clear. Firstly, we knew what we are doing. Secondly, we had a proven track record – if you come to Lansdowne there will be no problems with the sound; it will be good. And, if you want to degrade it for your particular needs, we will do that too. I make no apologies for this statement. It is not arrogance – just fact. Our pooled engineering experience meant we listened to a producer's requirement for a certain sound and then delivered it. We preferred to discuss the session beforehand to understand what is in mind. We were competent in knowing what people wanted if they described it. We never tried to run a session, unless asked: we gave the client what was asked for and never what we thought the client ought to have.

Denis Preston knew Bill Russo, who came to London from the US in 1962. Bill a trombonist, composer and arranger wrote charts for the Kenton band in the 50s. Denis told me that Bill had a rehearsal band and would be using Lansdowne as a rehearsal studio, the rehearsals would not interfere with the day-to-day running of the studio and in any case the band would be recording an album for Record Supervision. The rehearsals took place on Sunday afternoons or late at night when the band guys were free. It was a wonderful handpicked band of distinguished musicians. The recording approach was the minimum use of mics: the band would have a good internal balance. That's what Bill wanted; I asked Vic if he would like to record the rehearsals and eventually the Album. He agreed, it would be good to check the phasing of all the equipment on rehearsals. Here we were going back to room mics but not so distant from the band: the band was internally well balanced. The arrangements were originally written for the Kenton band in the 50s and reworked for the Lansdowne sessions. Something like half a dozen mics was utilised KM54s, SM2 and U67s. The recordings took place on December $21^{st} - 22^{nd}$ 1962. The band line up, excluding Bill, was four trumpets, five trombones, five saxophones, guitar, double bass, percussion and four cellos.

The studio setup had the rhythm section laid out in a line down the centre of the studio dividing the trumpets, trombones, saxophones and cellos in two making two smaller orchestras. One left one right. It worked very well and the phase coherence was excellent. The reissue *Russo in London* CD can be found on Vocalion records CDSML 8490. A bonus on that CD is the Kenny Baker's dozen recorded by Joe Meek 23rd 24th February 1959. First time to be released in stereo in which the Lansdowne echo chamber can clearly be heard on *Influential Character*.

In 1962, Keith Prowse Music, booked Lansdowne to record group audition demo records. I had worked previously with them in early '61 on Library music recordings with Ted Heath and Laurie Johnson.

1962 saw fast changes in music industry genres, especially in the popular (Pop) music area. Out went middle of the road records – the singer and orchestra type, some traditional Jazz, Skiffle and so on. Although these records have their place, in 1962 the group scene in the UK was being born – it started in 1954 in America with Bill Haley and the Comets' *Rock around the Clock* followed by Elvis Presley in 1956 with *Heartbreak Hotel*. These recordings marked a seminal shift and were a wake-up call to the record companies that audience's tastes were changing for the first time since World War II.

It appeared to me, as a 24-year-old, that many record company A&R people ended up auditioning any groups they thought would make it. We called the groups who came in for demos "three-guitars-drums-all sing"!

Amongst these groups were a number of Irish show bands, eager to make their name by capitalizing on the reputation at Lansdowne for making hits. Word spread like wildfire among the Irish, and we went through a period of recording Irish show-band groups. They would come in do their session, usually with a producer who didn't know his backside from his elbow sitting in the control room trying to look like he knew what he was talking about – usually rubbish. Session finished, band left, never to be heard of again. The so-called producer should never have given up his day job. Talk about ships passing in the night! We made sure we received our money up front!

Chapter 12

Robin Philips – Recording KPM Library Music Library, in Cologne, Munich and Brussels

Background

My association with Keith Prowse Music at Lansdowne goes back to 1960, recording demos and early library music for the "firm", Bill Phillips – Robin's uncle, Jimmy Phillips was MD and father of Robin and his elder brother Peter. I was audio engineering for KPM from early 1960 onwards. In those early days I recorded the Ted Heath Band and Laurie Johnson for the library with Bill Phillips in attendance. Then with Robin when he took on the library in 1965. Robin and I worked together on many projects for KPM Music Recorded Library (as it used to be known) for over 30 years and on occasions for Brother Peter.

The company was originally called Peter Maurice, a part of the Keith Prowse ticket agency. In 1955 Pat (Patrick) Howgill was responsible for the management of the library recordings issued on 78 rpm shellac discs then subsequently Desmond Irwin. In 1959, Associated-Rediffusion purchased the successful Peter Maurice publishing company and combined it with Keith Prowse Music Publishing to form KPM – Keith-Prowse-Maurice shortened to

KPM. There were two arms of the business. The music publishing was Keith Prowse Music Publishing Ltd and The Peter Maurice Music Company Ltd reflected on the headed notepaper. In the middle '60s Robin Philips was appointed to run the recorded music library of KPM, known as the KPM music group. In 1969 EMI bought the music group by acquiring 100% shares from Rediffusion Holdings and kept the name. KPM is now part of the Sony ATV Music Publishing.

The first sessions: May 1966

In early spring 1966 I had the usual jovial phone call from Robin; "Mr K", (as he always called me),

"We are going to Germany to record our library music – Cologne in West Germany. Are you up for it?"

In the 60s the cold war was at its height and sadly Germany was divided into East and West with a huge wall built by the Russians to shut the East off from the West – the Berlin Wall.

Just like that, out of the blue! Typical. It didn't take the blink of an eyelid to say "Yes". Then, after the call, all the questions went through my mind. Which studio? How big? What's the line up for the orchestra/s? Which composers? What equipment does the studio have? Is it any good? How many tracks? What are the musicians like? What about a good rhythm section? And so on. Robin replied, "Don't worry, Mr K, this is a venture into the unknown – but it'll be alright! It's a good, large studio just on the outskirts of Cologne." The big question was why. And that was it – a challenge! And it was, too. Although to call it a challenge was big understatement from Robin.

Robin explained KPM were going to record a new library with a different approach and "modern sounds". I pushed him further to explain.

"Modern arrangements with new composers writing different styles of music with large and small orchestras and a different approach to recording, clean sounds upfront and present (effective) when called for and instruments EQ'd where necessary, compositions tailored with the ability to edit out of the main piece, shortened versions 30 or 60 second cuts for TV Jingles and "stings" for TV (a sting is a short musical phrase known as a bumper of five seconds or less and acts as a form of punctuation to introduce a regular section of a show or dramatic climax). Robin had the foresight to anticipate what the market in library music wanted with fresh new recordings written and arranged by some of the best composers of their time: Johnny Pearson – well respected for his Top of The Pops Orchestra on BBC television and Syd Dale pianist and arranger, were the first to venture abroad with us. As it will be seen later, over the years, many other talented composers were invited to write for the library with great success.

In those 60s days there were no synthesizers, no high end processors and no solid state reverb. I used mostly a natural echo chamber for reverberation, sometimes EMT Plate and relying on the correct choice of microphones, their placement and very importantly the orchestral set up to maintain good separation between instruments (without the musicians losing eye contact with each other) and not least the recording console's first class equalisers and a good acoustic space.

Robin's brief gave me much to think about. I decided my approach to the recordings (partly from the pop school) should be fairly tight section miking on the big band and rhythm section, good separation on the strings from the rest of the band and, if possible, use the studio acoustic to help with perspective and "air" around the recordings. A different approach to each session line up over the years ranged from large orchestra of 60 with strings, including harp, to big band augmented with horns – usually 4 – to small band ensembles with string quartet with a small woodwind section and jazz influence pieces. This entailed having a discussion in the UK with the composers to see what was in their head about expected recorded sound and Robin for his brief, homework behind the desk again. This was my turn for adventure into unknown not ever having visited the studio at this stage. Over our time on all sessions Robin's pet saying was "Mr K – more 'air' required". To which I always responded, "No problem Robin! I have it cracked." The earliest track on CD KPM 643-6 is *The Trend-Setters* by Laurie Johnson, and this was used as the theme tune to the BBC's *Whickers World* until 1968; I recorded this in 1960 at Lansdowne.

Library music was from an earlier era, mostly orchestral light music composed by well known concert orchestra composers and used by the BBC, the country's main broadcaster pre-commercial TV and radio. Some well-known signature library light music, not all, KPM includes:

KPM1245 *Horses For Courses* composed by Paul Fenoulhet

KPM LP1245 *Highly Strung* composed by George French.

A commercial release Columbia DB 2595: *In Party Mood* by Charles Williams Concert Orchestra composed by Jack Strachey. The theme for the BBC's *Housewives' Choice* radio programme for many years.

A commercial release Columbia DB 2406: *Coronation Scot* (a musical depiction of a train journey) written by Vivian Ellis, and performed by the Queen's Hall Light Orchestra conducted by Sidney Torch, which was used as the introductory and closing theme music for the majority of the long-running BBC radio series, *Paul Temple*, from 1938.

Grandstand, titled *News Scoop* was used as the original 1958 theme tune composed by Herbert Leonard Stevens and it was one of the most iconic BBC TV themes of the 1960s, this theme was used from 1958 till 1971 when it was replaced by my recording for KPM written by Keith Mansfield.

Commercial release Columbia DB 2406: *Horse Guards – Whitehall* composed by Haydn Wood and performed by the Queens Hall Light Orchestra. This was the theme to long running BBC radio series *Down Your Way*.

Boosey and Hawkes Library music O.2083. *Barwick Green* the theme music to the long-running BBC Radio 4 soap opera *The Archers*. It is a maypole dance from the suite *My Native Heath*, written in 1924 by the Yorkshire composer Arthur Wood, and named after Barwick-in-Elmet.

Chappell Music Library: *Devil's Gallop*, performed by Charles Williams and his concert orchestra and composed by Charles Williams, better known as the theme tune to the radio serial *Dick Barton – Special Agent*.

Columbia DB 1945: *Calling All Workers* composed by Eric Coates and performed by Eric Coates and his Symphony Orchestra. This tune was the well-known theme to the long running BBC radio programme Music *While You Work*.

Listening to many of those recordings they were typically a classical approach to recording light orchestral using overhead room mics and much room acoustic sound in the recording. There was nothing wrong in that but times in the 60s were changing. A fresh approach was required. It was a very foresighted decision by Robin and Peter, and a risky investment on the part of KPM, that fortunately paid off for the composers and KPM over many years. Some say the '60s and '70s were iconic years for KPM recorded music library.

These recordings could, and should, have been done in the UK but the Musicians' Union made it too expensive. In the early 60's, the British Musician's Union in their doctrinaire "wisdom" effectively by restrictive practice debarred library recordings because of the expected high cost of multi-use which is the whole point of library music. The musicians had to be paid again wherever the work or edit pieces were re-used.

There was no buyout and freedom for the publisher to place the music without any burden of paying musicians again. An administrative nightmare to boot! The consequence was to record on the continent. Over the years of recording the UK session musicians lost work and consequently much money thanks to union intransigence. It also became clear to us that the MU did not understand changes in the music industry.

There was one library music owner/producer who circumvented the MU rules and worked recording at Lansdowne. He pretended (to the unions and others) the recordings were for a commercial album and not library music. Lansdowne used to have a break room where musicians would go for a coffee or tea and a fag. After one particular session had ended I went to see some of the musicians in this room and there was this producer dolling out bundles of cash to the writers. I commented to one writer that it was unusual to get paid for library music as they usually received the royalties from the PRS (Performing Rights Society) and MCPS (Mechanical Copyright Protection Society). This guy said, "We don't get any mechanicals but we get the PRS." Not realising of course it was not only PRS royalties but it was the mechanicals in library music that made them the money. How disingenuous of that publisher – a wolf in sheep's clothing.

Robin's approach was to take the work elsewhere. This was not in any way dismissive of British musicians as he felt, quite rightly, they were the best in the world, but a challenge to the MU officialdom who dictated this stance. The MU and its officers were steeped in old union ways from a bygone era – they did not understand the way the recording world was fast progressing and changing recording techniques. The intention to protect their members was good, but it had a deleterious effect on the musicians many of whom lost much work. It wasn't until the late '70s that Robin and others secured an agreement to work with London session players in UK studios with an affordable deal.

Musicians' union: *"The Union opposes their members recording as part of any television programmes which are to be recorded for repeat or subsequent broadcast."* Source: Union history 1950-1960. This quote sums it up! Library music is used for TV!!

I called Robin up and asked for the answers to my questions. The conversation went something like this;

RP – "We go over in June to a studio called Ariola in Cologne" (corporate name Sonopress Tonstudiotechnik GmbH who owned two other recording studios, Berlin and Gütersloh).

AK -"And microphones?"

RP – "I am told they have plenty of good ones"

AK -"What exactly?"

RP – "Not sure but the recording equipment is new".

By now I am thinking, what I am going to walk into at Ariola?

RP – "Mr K it's alright. I can assure you I've spoken to them"!

AK – "OK Robin. What are the musicians like?"

RP – "Well for a start we are taking a rhythm section and a lead violinist"

AK – "Who exactly?"

RP – "Kenny Clare on drums, Peter McGurk on bass (acoustic), Alan Parker on guitars, Jim Lawless on percussion and Tony Gilbert (violin) to lead the strings."

All hugely talented and experienced London session-scene players. These were superb players I knew personally, having worked with them at Lansdowne on many sessions, and so I was reassured.

AK – "What's the line up?"

RP – "Musicians from the Kurt Edelhagen Big Band and the Kenny Clarke/Francy Boland big band augmented with the Cologne symphony strings – to be led by Tony Gilbert."

Members of the two bands formed the Westdeutscher Rundfunk (WDR) Big Band. That was good enough for me – it was some line up! Background: The WDR Big Band was formerly called WDR later becoming, with complete new personnel, the WDR Big Band. The Edelhagen band broke up at end of 1972

AK – "Who are the composers?"

RP – "The scores will be written by Syd Dale and Johnny Pearson."

Both these guys were excellent composers/arrangers in their own right and had been carefully briefed on the genre of music each had to compose. I knew and respected Syd having worked previously with him at Lansdowne and we had built up a good rapport. I knew of Johnny Pearson but had never met him, as at that time he was recording at Pye Studios Marble Arch and had hits with Sounds Orchestral (John Schroder – Producer). He was a magnificent pianist and composer who studied piano under Solomon.

Robin had one more card up his sleeve,

"Each session will be four hours in length and we have plenty to do! This is a bit of an experiment to see how things work out."

No pressure then! It was a challenge I looked forward to with some trepidation; it certainly was an adventure into the unknown. Robin called me up again to suggest a departure date and to let me know that his brother Peter was joining us. A date was scheduled for May 1966!

"Robin," I protested, "I need to go out a day earlier to see the studio, get a feel for the acoustics, to set it up and to know who I am working with."

Thus I found myself on a BEA plane from Heathrow to Cologne a 727 Boeing intercity jet, a day before everyone else. Bookings for all of us were made at The Königshof Hotel near the towering Gothic Cologne Cathedral and Hohe Straße, incidentally one of the city's oldest and very busy pedestrian shopping streets rebuilt after the war. During WWII Cologne was carpet bombed – it was a miracle the cathedral survived. I had seen pictures of the devastation Cologne had suffered in the war with the 1000 bomber raids over the city. What particularly struck me was the contrast between the rebuilt Hohe Straße and the shell-pockmarked cathedral – the stone exterior of which was very much in need of cleaning and repair. At 11DM to the pound, goods in shops and restaurants were exceptionally good value for money. The rest of Cologne had been rebuilt with brand new trams running and first-class wide roads.

Ariola studios were located in Unter Kirschen 8, a street in Cologne suburb called Bickendorf. I was warmly greeted by two guys, Manfred Kabunder, who was the technical engineer and a younger guy than me, Justus Liebich, who was to be my assistant. It was a studio larger than I had imagined and acoustically live. It was more than able to accommodate the varying size of orchestras we had to record – from small groups to 60 plus. I needn't have worried. The studio volume was 1200m³ with a rhythm section separation area raised off the studio floor and a low roof 250 m² in area. There was a relatively new Steinway B Grand Piano. I had a good choice of Neumann and Schoeps condenser valve (tube) mics and some dynamics – more than I required. The control room had a new twenty-channel input Siemens console with comprehensive channel equalisation and pan pots (panning potentiometers). There were faders I had never seen before called Danner faders – flat with a long stroke for accurate level control, so much better than the Painton noisy stud-quadrant type we had at Lansdowne, which accumulated fag ash! Danner faders were better protected. There were excellent Altec 604 monitor speakers driven by Telefunken V69a valve (tube) amplifiers, which were very accurate to record and mix on. They were great sounding monitors. The kit also included two Telefunken M10 4/6 multi-track machines. The multi-track was 1″ six track, only being able to record on four tracks simultaneously there was no meters located in the machine to monitor the individual tracks. The other two tracks (5 & 6) for overdubbing required switching the record amps to the other two tracks (there were six playback amps) – I thought it most odd but when needed it worked. Two other Telefunken machines were stereo/mono compatible M10A fitted with butterfly heads (0.75mm head gap not the usual 2mm gap). For the not so technical, it meant that the stereo recording (0.75mm head gap) could be played back on a mono machine with no change of audio level and in mono, no need to make a mono master. At that period of time Lansdowne didn't have that type of head; the Ampex recorders had 2mm head gaps only. When we bought Telefunken M10As in the early '70s 0.75mm butterfly heads came as standard. No more losing sonic quality through collapsing to mono. It is fair to say in that '60s period UK studios didn't "get it" when I insisted if a master lacquer had to cut mono, just play back the stereo and you have perfect mono with no sonic copying loss. Ariola had an excellent limiter, the Fairchild 670, which I was at home with as we had one at Lansdowne.

The reverberation was two EMT plates and an excellent natural echo chamber (called Hallraum – fixed 3.5 second reverberation). The studio was well maintained with excellent German efficiency and I had fabulous cooperation from all the studio staff. Nothing was too much trouble. To listen for perspective on the recording to the four tracks (dry) I used a few milliseconds delay to the chamber. The delay was achieved by sending a chosen channel auxiliary send to the record input then the output (of the machine) to one track of the Stereo ¼" M10 Telefunken and using vari speed to control the delay. The chamber return was to the monitor mix only. Robins "air around the sound!" This was an addition to making the room acoustics work for me.

The recordings were planned to capture everything live and with atmosphere. The performance is so much better that way. I made good use of the natural chamber and EMT plates using a combination of them on the mixes. There are many examples in my recordings. However, a good example of combination reverb can be well heard on Johnny Pearson's "Heavy Action (recorded later in Decca Fonior Studios 1974) - a famous introduction to America's Monday Night Football and used for many years as the theme to UK Superstars.

In that band line up I always liked to work with the musicians not having to wear cans (headphones) unless absolutely necessary. The aim was to avoid the use of overdubs in possible therefore the studio layout was critical so the guys could hear each other. The exception to the rule was the rhythm section – they needed to hear a selective foldback mix of the whole band or sections as required. The rhythm section was on a raised booth area on the back wall of the booth and facing out to the studio floor so they had line of sight with low permanent screening in front to maintain good separation – especially from the close-by string section.

The guys asked me about the orchestra size for the studio set-up – it was our four rhythm guys (drums, acoustic bass, guitar doubling electric guitar, percussion and piano, four trumpets, four trombones, four horns, five saxes, twelve first violins, six seconds, four violas, four celli and orchestral bass (on some pieces). We spent about three hours setting up; I then had to get my head around this unfamiliar Siemens console about which I knew nothing. I needn't have worried. Justus was very much on the case, thank goodness!

There was no patch field as we know it but an unusual matrix of six pin sockets with separate mating plugs, inputs and outputs on the same plug therefore one multi patch chord only, located either side of the peak level meter for track/group selection to the machines, for reverb, outboard gear and so on. The channel equalisation was most comprehensive. The audio level peak meter was a *Müller & Weigert* galvanometer light beam Peak Programme Meter (PPM) typically used by German broadcasters so much better to use, more accurate with expanded scale, than the old, small, round-faced BBC type.

We checked out that all was working; the mics plugged into the correct channels, identified and grouped to the recorder tracks and working for the first session next day. The session with this orchestra was due to start at 9.00am. No such luck! At 9.00am our guys were ready to go as was the practice in the UK but large sections of the orchestra were missing. They came in dribs and drabs and by 9.30am we almost had a complete orchestra. The brass section was the last to arrive. Swiftly followed, to my total surprise, by a waiter carrying a tray of breakfast delivered to some of the guys in the trumpet and trombone section. This seemed to be a ritual before the start of every morning session. It appeared the customary breakfast food was "strammer max" (ham and eggs on fried bread), a hangover cure for the night before! An everlasting memory was that of the guys in the trumpet section using their instrument cases as tables!

It transpired the waiter came from the bar/restaurant behind the studio that could be accessed via an interconnecting door. Oh disaster, I thought. A bar and musicians in such close proximity doesn't always work especially with thirsty brass players in need of "tea" breaks. I needn't have worried though as it worked out well; we used that bar for a few Kölsch beers (the local Cologne brew) after a hard day of recording.

Robin and Peter were getting most anxious about the late start, as there was much to record. I learnt later Peter came to keep an eye on his younger brother for these first sessions; this was a very expensive exercise.

On the first day of recording Syd Dale was on the conductor's stand poised to start after having introduced himself to the musos. I asked for a run through to get a balance. Syd's count-ins were "a-one-and-a-two-and-a-three-a and… bosh!" The band played. The piece was loud and complicated to play. The PPM meter went way into the red, off the end scale – ouch! Some of the players had not experienced such a score, chiefly the classical string section. Tony Gilbert, lead violinist worked hard to get the strings up to speed. Fortunately there were rehearsals, which gave me time to get some decent section sounds, particularly with the rhythm section as the backbone of the score. For me, it meant working with a strange recording console I had never seen or worked with before, with a peak reading meter I also had no knowledge of (only BBC-type PPM), but thanks to the rehearsals I got my levels under control – the meter was edging into the red but experience told me this was tolerable. I had to very careful about where to put which orchestral sections on which tracks as I only had four tracks to record on. On this first session we had rhythm and percussion, strings, brass to include horns and saxes. In string-based pieces, I placed the strings in stereo across two tracks. Having only four tracks I changed the track layout for each composition, if required. This then set out the basis of the four-track recordings. My philosophy, therefore, was to monitor the mix in stereo and balance the band as if I were recording direct to stereo (I had plenty of experience at Lansdowne when, in the early days all Preston's material was direct to stereo). There was no room for a margin of error in the balance: for example the rhythm section and percussion (on some pieces) was recorded to one track and had to sit well within the overall balance. It had to be right first time. However it gave us flexibility for the mix.

The twin light beam meter summed the outputs of all tracks left and right and went well into the red most of the time to the amazement of Justus and Manfred.

"Don't you think the level is too much?" they said.

"No," I said, "it's fine. We are recording to Agfa 555 and this meter is showing the summed console track outputs.

Still they questioned me,

"We don't record this much into the red".

It appeared it was frowned upon to record into the red – you can't do that! I didn't have time to explain how the summing added to the metering level. Unfortunately it was not possible to switch the meter to read the individual track levels on record. Later when I looked at the level on individual tracks they were recorded "hot" to an indicated level of +10dBu. No tape saturation – well it didn't sound like it! The recording across all four tracks was clean.

Some of the string players, being classical musicians, had never heard or played arrangements like it so it took the band some time to get into the way we recorded in London. There was also a communication and language problem with some of the string players and other members of the orchestra. This was overcome by a translator who was himself a bass player, John Fischer. The Big Band was fabulous, consisting of a few luminary Jazz players in their own right:

Trumpets featured Milo Pavlović, Duško Gojković, Rick Kiefer, Shake Keane (whom I knew and worked with in London when he recorded with Joe Harriott for Preston)

Trombones: Jiggs Whigham lead, Otto Bredl, Manfred Gaetjens, Nick Hauck.

Saxophones: Derek Humble (alto), Heinz Kretchmar (alto/clarinet), Karl Drewo (tenor/alto), Wilton Gaynair (tenor) from Jamaica, Kurt Aderholt (baritone).

On piano was Francis Coppieters except when Johnny Pearson played piano on a number of his own compositions. Since the big band played together regularly they really swung, and were fabulous to listen to. And finally, not forgetting the four horn section, players from Cologne Symphony Orchestra. Horns were not used on every session. They were a bunch of hugely talented musicians. Members of the Clarke/Boland Big Band knew Kenny Clare who played with the band on a regular basis, commuting from the UK. The synergy between drums and this studio amalgamation of a Big Band was remarkable. Professor Jiggs Wigham comments on the sessions: *"We nearly all smoked (cigarettes)!! We often did only one take. The Germans did, and still do, love, to do many! Listening to Kenny Clare (we became close friends) was a joy! Tony Gilbert (violin) too and that you loved to record 'hot!'"*

I struggled with the instrument separation, particularly with separating the strings from most of the band with too much bleed into the string mics. We did record the four hours, sometimes a little over – no union rules about overtime to hinder us here! Session hours were quite relaxed and the entire band was most cooperative. I, however, wasn't terribly pleased with the result of the morning's work. The four-hour afternoon session, same set up, with Syd was easier on the band and myself. Although the recording was acceptable to Robin, I knew I could do better. I said to Robin,

"I have to improve the separation of the strings from the Big Band for tomorrow and change some of the set-up. Leave it with me as this is not the true result required or discussed."

"What are you going to do?" he asked.

My idea was to build a canopy over the string section to reduce the studio ceiling height in that area, hang a lightweight drape material, which the studio had, on the wall behind the section and a booth for Jim to stop the percussion sounds from flying all over the studio. It wasn't a disastrous recorded sound but I wanted controlled separation to give depth and width. As Robin called it, "air around the sound", with punchy up-front rhythm. Fortunately the drums, bass and guitar were elevated (from the studio floor) under a low roof at the side of the room. During the sessions Peter sat in an adjacent room listening through an open door sometimes passing comment. He said the monitoring sound in the control room was too loud, I thought he was being diplomatic to let Robin get on with producing without Peter looking over his shoulder.

I asked Manfred and Justus if they had some material to build the canopy for the string section, the answer was no but they said they could get something delivered. What turned up was a large builder's sheet. I had enough movable screens. We worked late into the night erecting it and a percussion "booth" with studio screens and a temporary roof then resetting the studio. I was pretty shattered! Anticipating this I asked Robin if we could start the session at 11.00am. He asked the fixer – Ferdy Klein to arrange it. The following day there were two sessions with Johnny Pearson. I had complete control and a tighter sound – wonderful, I had good separation on the string section and good control. This was more like it! I got to know the guys in the band pretty well and built up a rapport with them, they heard my recorded sound on playback and clearly trusted me, one commented, "Who is this English guy who this gets this good sound?"

On the previous day, under the console, my right foot contacted what I thought was a protruding lump. I gave it no more thought and didn't bother to peer underneath. I dismissed it as part of the console construction. During the second morning session, Manfred and another guy in a white coat came into the control room carrying a large object. They asked to stop the session – *what?!*

Robin: "What's happening Mr K?"

I wasn't sure. The guy in the white coat was carrying a volt meter, he bent down under the right hand side of the console and made a measurement, got back up and said, "OK we have to change the battery!" I was lost for words – this is unbelievable! Battery duly changed, the session resumed. It was a heavy-duty 24-volt battery: the transistorised electronics worked with +24 volts DC. I thought I must get to the bottom of this. Robin was amused, I was unnerved. At session end I asked Manfred what was going on.

He explained "This console is brand new it and was only installed the day before you arrived and you are the first to use it. We had much trouble with mains hum, which we think is something to do with the power supplies, so we are using battery power until we resolve the problem". So I was the guinea pig! No wonder people kept popping in and out of the control room to listen to the sound including a very well dressed guy in a suit, Herr Schulze – the studio manager. He was extremely correct, very polite and when we shook hands he clicked his heels and gave a small bow.

No trial session then to sort out any wiring bugs! I was it! Fortunately there were none and I didn't hear any potential phasing problems. Periodically the same guy in the white coat would pop in with his volt meter to be sure the 24 volts had not dropped too much. I had no idea of the amps' drain on the battery. I thought it best to leave it to them while I get on with the recordings. The second day, Johnny Pearson was on the conductor's stand. The arrangements were brilliant and the recording was easy now I had solved the string separation problem, and the band had got used to our London way of working. This was the first time I had worked with Johnny and what a joy it was! After the two days recording I felt we had a good result – fingers crossed for mixing back in London.

For the sixteen hours of recording the pressure was really on me. Frankly, it was a kick-bollock-and-scramble at times. Always to the strains of Robin's "Come on Mr K, there's a lot to do! We are behind!" There were no playbacks in the box, except the first title of the day to let the musicians hear the results of their endeavour. This was to be the norm on all sessions as Robin insisted "Mr K we don't have time to play each number back. We have to crack on."

231

Only once did Robin question my decision: "Now Mr K, what are you doing?"

"I am changing some track layouts for this title – as I did on many other titles – we only have four tracks to work with and I want the best configuration for the mixing sessions."

After that I was not questioned.

If there were performance edits needed I asked for them but Robin didn't always agree. His response was, "It'll be alright – we'll fix it"

Sometimes I had to be more insistent,

"No Robin there are duff notes and it's not a good performance – we can't fix that in the mix. Let's make an edit and also choose between the takes for the best performance. For example let's make the edit section two bars before letter F to two after I, that'll work, we'll cut from F to I on the mix." With this discussion going on in the box the composer would ask "what's going on can I move on" – they felt the session pressure as well. It was down to me to explain and suggest the edit section/s

Edit sections were always agreed with the composer/conductor as they didn't always hear the mistakes in their cans.

After the two days of recording the idea was to go back to London to mix stereo masters from the multi-track tapes. There were no problems getting the tapes back through customs. No common market then! No problem doing business with Europe and I never experienced any in the '60s to mid '70s, when I periodically worked in Germany and Brussels.

Lansdowne did not have the capability to playback 1" four track – just ½" four-track. Robin had booked Levy Sound in Bond Street and his brother Peter was with us. Levy Sound was a disaster. There was no console equalisation just a couple of Pultec equalisers (nothing wrong in that but no good to us), no good limiter compressor, dreadful noisy quadrant faders, lousy monitoring, no assistant, a 1" four-track not six as in Ariola. In essence the track alignment was out, something I explained to Robin to no avail, fortunately I had not used the other two of the six tracks in Cologne. We were on our own.

In short the studio was not a set-up for the type of work we were doing. It was more like a mastering and programme production studio. I didn't even get round to checking out the reverb. The whole shebang needed good maintenance. After spending about an hour struggling to get a result and no assistant tape operator I told Robin it wasn't going to work. He called up Jacques Levy and explained the situation in his own incomparable way then passed the phone to me. I guess I was going to be the fall guy! In the most diplomatic terms I could muster I explained about the equipment. Mr Levy was decidedly very discontented that I had criticised his studio but had to accept the situation. That was it then! Phew!

That left us with another problem.

"What are we going to do?" asked Robin.

My solution was to go back to Cologne to mix as they had all the equipment we required.

"Right," said Robin, "We go back next week. I'll organise it with the studio."

"OK but it's Friday evening and the studio may not be available."

Robin was adamant,

"I'll get the money and book the studio and flights."

And Robin made it happen. To obtain the money Robin had to get into the KPM offices, which he did without any keys. I leave it up to your imagination how! We returned to Cologne on the Tuesday, lugging with us the multi-track masters (no extra baggage charge!) with explicit instructions not to place the tapes near any electrical equipment in the hold. Again customs were not a problem at Heathrow or Cologne – this was the years before security screening and scanners. Back in at Ariola, the mixing was a doddle. I used a Fairchild 670 limiter on the overall mix just to hold music peaks. Mixing from the four tracks was done, after all the recording sessions were completed, to Stereo/Mono-compatible Telefunken M 10 fitted with butterfly heads (.75mm gap). The tape we used was Agfa 525, which could handle very high levels. Not all tracks were recorded in Cologne. Later, over the years, a number of sessions were recorded in Belgium and Munich with some changes in the rhythm section. Interestingly, the valve microphone types used then are still around today and very much sought-after.

The recordings typify the music of the time and the approach to recording, now some 48 years ago, using the simplest of technology with no signal processors other than eq. We had excellent console equalisation and a fabulous echo chamber. As mentioned the main monitors were very accurate as were the control room acoustics. I always have been a believer in checking mixes on small speakers more aligned to a radio or TV speaker – I called them grot boxes: an apt name. However, in this control room I was able to low power transmit (AM) to a valve radio receiver in the room as another check on the balance to hear how the sound was on transmission and make any small balance adjustments if required. The stereo masters would be mastered for vinyl pressing – no CDs fifty years ago. I had the advantage of looking at a simulated RIAA curve (Recording Industry Association of America) cutting current meter and max current red line not to be crossed. An explanation: The master lacquer-cutting process requires equalisation; it employs the RIAA curve. Without trying to bamboozle the non-techies, in simple terms the process requires a 20dB bass cut and a 20dB treble boost when cutting a master lacquer. The 20dB treble boost can put excessive stress on the cutting amplifiers; circuit breakers are used in the recording chain to avoid burning out the very expensive cutter head. This high frequency emphasis also causes fundamental high-frequency instruments like cymbals and brass (a good case in point being Harmon muted trumpets) to peak the current because of the high-frequency content in the sound. This causes distortion if cut without care. One practical example when mixing the KPM material was if I had a solo harmon muted trumpet and placed it in the mix to where I thought was in a good perspective the cutting current meter would breach the red line, if not corrected this would cause the cutting engineer to reduce the overall cut level – didn't want that. But by reducing that muted trumpet level by a small amount only 2dB (which you don't hear in a complex mix) the problem was solved and the cutting engineer can maintain a good level to the master lacquer without grief. When the vinyl disc is played back the playback curve is converse to the recording curve.

We completed the mixing in good time over a period of two normal working days leaving the evenings free for eating out at good Cologne restaurants and to relax, courtesy of Robin. After two days we caught a flight back to London with a whole bundle of stereo ¼″ masters mixed to Agfa PER 525 leaving the multi-tracks at Ariola. Once again we trundled through customs without a second glance. It was more of a problem to again ensure the masters were stowed in the hold away from magnetic fields. The boxes were well labelled with strict instructions. BEA was once again careful and we never had any problems although Robin was nervous about them in the hold.

These initial recordings set the template for further future work (hopefully) we would do in that studio. I got to know the studio floor idiosyncrasies and the studio guys very well. Justus had first-rate musical ears and was very quick at editing the stereo masters. The musicians were outstanding and appreciated our way of recording always under pressure to get as much in the can as we could. Robin would sit at my right-hand side and give me the musical cues – for solos etc.

This, in a musically adventurous way, was the start of the KPM "new" music library and continued for many years with a massive amount of recordings. We recorded in Germany every year from 1966 to 1969. After the first '66 sessions Robin and Peter took time out, sensibly, to assess results and where the library was going for the future. We would next return to record at Ariola in 1967 and latterly Trixi in Munich – more about that later! Although the bulk of recordings were in Cologne we also worked in Munich and Brussels stretching into the mid '70s. The recording periods were generally but not exceptionally in late May or mid-June and for up to two weeks allowing for recording then mixing. A massive subject and a huge amount of recording – a subject in its own right well beyond the scope of this book. Those KPM recording episodes are worth a book in themselves. At the time of writing this, I understand a detailed KPM music library book is being written about that period.

I pay tribute to Robin and Peter for their far sightedness in creating this new library. It was hugely successful and I was very proud to be the chosen engineer on all sessions and eventually recording at Lansdowne. These early beginnings were the start of many pleasant journeys to Europe – including the early morning plane journeys back to London nursing a huge hangover (too many glugs of wine and beer!) from the night before. And then there were the early morning champagne breakfasts on the aircraft we enjoyed, to the amazement of the BEA cabin crew, a pick-me up after the night before to get us going. Robin did it in style! Our musicians flew back to London when the recording sessions were completed, leaving us to mix.

In the middle '70s, by negotiation, Robin and others managed to get the union to see sense and an agreement was reached for library recording in the UK at sensible buyout rates. Robin's approach has since been shown to be correct (library music recording is now embraced within the BMU/BPI Agreement), but it shows yet another side of his challenge of the unusual that put library music (and KPM in particular), firmly on the map. The economic loss to the London session players was substantial. Over the years it cost British musicians in lost work many thousands of pounds. I lay the blame firmly at the feet of the Musicians' Union. It was their short-sightedness and intransigence and above all their ignorance of the rapid advancing technology changes in the recording scene.

My dear friend Robin passed away on 13th May 2006 and dear friend Peter, on 10th February 2015. They were great times guys, never to be forgotten!

KPM was and still is a hugely successful music library and I was delighted to work with so many talented composers: Brian Bennett, John Cameron, James Clarke, Syd Dale, Johnny Dankworth, John Fiddy, Steve Gray, Alan Hawkshaw, Tony Kinsey, Dave Lindup, Keith Mansfield, Alan Parker, Johnny Pearson, Neil Richardson, John Scott, John Shakespeare, Bill le Sage, and Stan Tracey. Not all these composers flew out to Germany (there weren't sufficient of their compositions to warrant it), their compositions were conducted by Syd Dale, Johnny Pearson or Neil Richardson. It became mostly the norm for us to travel to Germany with Robin in May/June. Unbeknown to me at the time (1966), this was the early beginning of seven years recording in Germany, Ariola, Cologne – 1966, 1967, 1968, 1969, March 1969 Trixie – aborted session Munich. Then Arco Studios, Munich (later in the week) to replace aborted Trixie session. After 1970 came the Belgium sessions: Katy Studios 1972, Cornet Studios Cologne 1972, Fonior Studios, March 1974 and Morgan Studios Brussels June 1975. Then at last in 1976 we were able to record in the UK without MU restrictions.

Ariola, Cologne, May, 1966

As I have mentioned the sessions typically lasted four hours except the first day when the band was getting used to the way we worked, about half-hour overtime and exceptionally laid back. But in the box it was a different story: Robin would say,

"Come on Mr K there's a lot to do. What's happening? Where are the guys?"

"Well," I'd reply, "The guys are taking a break!"

"Why?"

"It's a "tea" break" (even though some players only drank coffee or alcohol, never tea) I don't know, they've gone to the bar behind the studio; I guess they've gone to get a drink!" was my typical answer. If they weren't at the bar a waiter would bring the refreshments in... Big band brass players were always thirsty!!

When we broke for the musicians to eat you'd hear Robin again, "Come on Mr K! We have to get them back, we have a lot to do". So I did with the help of Eustace as I spoke little German. Most of the Big Band guys spoke English but many of the string section did not. I had a good rapport with the rhythm section and there was much banter between myself and Peter McGurk between takes. On a technical note he always carried with him a Lavalier microphone (the big old hang-around-your neck type, a precursor to the small lapel radio mics in use today), which I used to record his bass wrapped in thick sponge rubber inserted between the bridge and body of the bass with the head facing up and some EQ lift at 1.4kHz. It produced a superb sound. It was an excellent acoustic bass to record. I first worked with Peter when I recorded the Dudley Moore trio. Kenny Clare, Alan Parker, Jim Lawless and Tony Gilbert were very much in demand London "A list" session men with whom I regularly worked on sessions at Lansdowne.

When I first started to mix at Ariola there were recurrent distant rumblings coming up from the control room floor then faint clicks on the monitor speakers. When I asked what the heck it was Eustace replied, "Don't worry. There is a bowling alley in the basement and the clicks are from the electrical mechanism lifting the skittles, it's not going to tape". I wanted to check and sure enough the clicks weren't recorded. Fortunately it wasn't very regular and the tech guys had resolved the problem for the next date.

As mentioned previously the studio manager was a Herr Schultze, probably in his late 30s, a charming man and very correct – I never knew his Christian name. He was always immaculately dressed and would visit the control room to see how we were doing. He would shake hands with a slight bow, click his heels and make complimentary comments about the sound then leave. This routine was almost on a daily basis at about the same time in the morning – a charming man. Every day the studio staff lunches were delivered, the strong smell of sauerkraut and coffee you could stand a spoon in always pervaded the studio corridors. Ever since then when I saw sauerkraut on a restaurant menu I was totally put off.

After the end of sessions, I frequently went to the bar/restaurant at the back of the studio for a couple of Kölsch beers to unwind with Manfred, Eustace, Robin, Graham Walker and some of the guys from the band. On one occasion there were four old scruffy looking German men playing dominos obviously having had a few drinks, we were sitting at a table near them. We had only been there for a short while when suddenly Manfred said we should leave immediately as they were talking about the "Englanders" and the war and it could turn nasty. We hurriedly exited and I wondered if they were left over Nazis!

Ariola, Cologne, June 1967

I recorded a large range of different composers' work with different orchestra line-ups – small groups and large – different genres of composition: the usual huge amount of work to record then mix at the completion of all dates – my preferred option. The two prime conductors Syd Dale and Johnny Pearson recording their compositions and conducting other composers' work. By now Justus and I had the studio set-ups cracked and the sessions were much more relaxed; the musicians knew what they were in for. I recall Johnny Pearson had written a marvellous suite – *Twentieth Century Portrait Suite* – which was subsequently edited into separate pieces, one piece named *The Awakening*. Within it was chosen an edit section that became the original *News at Ten*. This can be heard today, the same tune but a different arrangement on ITN news today, forty-nine years on from the original composition recording. On the 1967 recording Alan Parker's electric guitar can be heard clearly on this piece, another talented player to have in the rhythm section.

During recording sessions Robin was his usual cheery self: "Come on Mr K lots to do!"

Requests for a tea break were often met with the response, "One more title then we'll have it." The break was called and a number of musicians went to the bar for the "tea" break. Others went outside in the tree-fringed yard to smoke and get out of the hot studio. Some played Frisbee. I had never seen a flying disc before and was told this was a new sport imported from the USA. Jiggs and the other guys played incessantly during breaks – it was good exercise. I recall May and June were always hot and humid months – the studio more so as there was no air conditioning. The sessions held in May took about ten days to record and mix. This time it was more relaxed with no late night session instead we ate out at some excellent restaurants and bars. Robin was most hospitable. Syd Dale and some of the big band brass players propped up the bar many an evening opposite WDR Studios downing Kölsch beer at times to the early hours, I occasionally joined them conscious of the fact that I always had early starts and needed a clear head!

Ariola, Cologne, KPM September 1967 circa 12th September

Sessions recording more compositions for the KPM library to include those written and conducted by James Clarke with string quartet, Pete McGurk (acoustic bass), Kenny Clare (drums) and Francis Coppieters (Boland Big Band) pianist. Syd Dale with a larger orchestra. By now we had a good routine and the larger band was used to our way of working.

Ariola, Cologne, 1968

This was a very busy year for me recording at Ariola. Word had spread to the recorded library music publishers we were achieving good results recording library music in Cologne and that the studio and musician costs were acceptable. By then Syd Dale and I had set up a company, Motive Music Ltd, to record for other libraries. Syd plus other composers wrote the scores; Syd was to conduct and I was to take care of all technical matters and produce the recordings. 1968 went something like this;

KPM, January 1968: Record & Mix

Graham Walker joined KPM for his first sessions working with Robin. This was a two-day recording trip with Johnny to specifically re-record *News at Ten*. The original *News at Ten*, edited from a movement called *The Awakening* part of Johnny's marvellous work *Twentieth Century Portrait Suite* and using the main theme and an edited end – but that left the section unresolved. ITN required it to end with a long repeated tail with the chimes of Big Ben dubbed on back in London. Apart from that I had to copy the exact sound from the 1967 recording. Fortunately we had the same band and rhythm section and I had the record of the studio set up and mics used. Aside from recording the new version we took the opportunity to record more library music as Johnny always played piano on his own compositions. We spent the evenings relaxing at excellent restaurants with some good wine!

Syd Dale, March 1968 Motive Music for Wienbergers: Recording and Mixing

For Stephie Lengauer head of Weinberger' recorded music library.

Two seven-hour record sessions and two eight-hour mix sessions with 12 musicians, totalling 30 hours. The charming Stephie, who ran the Wienberger music library, approached Syd Dale to organise the trip to Cologne to record compositions by Syd and Johnny Hawksworth. They were released on the Impress label owned by Weinberger's. Once again we used musicians from the Kurt Edelhagen Big Band and our usual rhythm section minus Kenny Clare who was booked on another date in the UK. The sessions were excellent by now as the musicians had played on our previous sessions and knew what to expect from English composers.

Line up:

Piano: Bora Rokovic. Drums: Toni Inzolako: Acoustic Bass: Pete McGurk. Guitars: Alan Parker. Percussion: Kurt Becker. Trumpets" Milo Pavlovic & Shake Keane. Trombone: Jiggs Wigham. Tenor sax: Karl Drewo. Baritone sax: Kurt Alderholdt. Marimba/Vibes: Claudio Szenkar.

Syd Dale, May 1968 Motive Music for Synchrofox Music Library Ltd AKA Sam Fox Productions

8th and 9th eleven and a half hours recording with 14 musicians plus Kenny Clare (drums), Pete McGurk (acoustic bass), Alan Parker (acoustic/electric guitars). 10th May, ten hours mixing. We recorded 25 titles (50 minutes of music). The piano player was Francis Coppieters, a marvellous musician who played Hammond organ on some dates in 1968. From memory, the recordings were released across two LPs. SF1017, SF1018: Composers: Syd Dale, Denis Wilson,

Malcolm Lockyer, Ronnie Hazlehurst, Brian Fahey, Andrew Arvin, Alan Roper – with compositions of all various musical genres. It was another successful Ariola session.

The musician line up: 2 flutes, oboe, alto sax, tenor sax, horn, 3 trumpets, 2 trombones, piano, drums, percussion, acoustic bass, acoustic/electric guitar.

KPM June 1968

9th to 19th – recorded and mixed for a total of circa106 hours, not including rest and travel days.

These dates were the largest Robin booked and the orchestra line-ups ranged from string quartets with a rhythm section to large orchestras. The main composers/conductors on the dates were Syd Dale, Johnny Pearson and Keith Mansfield and other composers' compositions who were not attending the sessions. So many titles were recorded it is impossible to list them all suffice to say that the recordings and mixing was most successful. I was shattered after those long days!

On one of the recording dates Johnny Pearson had a large number of scores to record. When he arrived at the studio he looked a little tired. "Are you all right Johnny?" I asked. "Oh I'm fine. Been up half the night checking the parts for copying mistakes".

In 1968 there were no home computers. The scores and parts were brought to the UK by each composer. All score parts had to be individually copied note by note, in ink, by specialist copyists. Their calligraphy skills were something to behold. The session pressure ("Much to do Mr K!") and the need to get as much in the can as we could meant any mistakes would hold up the session and momentum would be lost (see chapter 11 page 199 Frank Chacksfield session.)

We had some great times on those dates and a sad one: Peter McGurk took his own life back in London in June 1968. He flew back to London one morning before the sessions were completed, devastated by the breakdown of his relationship with his girlfriend with whom he was besotted. She had broken up with him the night before, at the Königshof Hotel.

Peter McGurk's tragic death is one that still, 48 years on, makes me stop and think. On sessions he was my jovial banter friend from my talkback to his cans (headphones). I had worked at Lansdowne with the Dudley Moore Trio with Pete on bass and Chris Karan on drums. He was a marvellous musician and wonderful bass player. A big loss for our sessions and the other guys. I wrote a personal letter to Pete's mother expressing my deepest sympathy and regret at his passing – I had a moving reply from her.

At that time we had more material to record and used bass guitar to complete the date, Peter Trunk was booked for that date; a player who also played acoustic bass. We needed a permanent replacement for future recordings and Dave Richmond filled that seat. Dave also played double bass and was affectionately known as the Player of Basses. After the dates Robin ever mindful of cost sent me the following letter. I managed to have the studio cost reduced as I was engineering, the studio did not have pay for their balance engineering in-house service.

Munich 1969
Trixi/Arco Studios, March 1969, Neil Richardson

Our first trip to Munich to record music composed by Neil Richardson was a disaster! We flew to Munich for sessions at Trixi to record the compositions. I set up the studio the night before and was ready to record promptly at 10.00 the following morning. At 10.00 a couple of musicians arrived, the rest trickled in in dribs and drabs, Robin was not best pleased but we managed to get started at about 11.00am. Neil's compositions were always well orchestrated and not difficult to play for competent musicians but the session was a total nightmare and Robin abandoned it less than an hour after we started. The playing was substandard and the equipment inadequate. The orchestra was simply not up to it, being straight (classical) German musicians didn't grasp the intention/interpretation of the writing. The strings were not good either. The session was a write-off and I am not sure whether Robin paid them. The players couldn't hack it although they were supposed to be the cream of Munich musicians. They fell far short of the mark. I told Robin this was not the standard we had not set out to achieve, nor had achieved in Cologne – he concurred. The recording console was a badly maintained MCI with a number of faults, making my job incredibly difficult.

Robin's KPM sub publishing representative had booked a band through a fixer who hadn't worked for 20 years. Needless to say, Robin had words with the fixer.

"We need a better band," he said. "We have to cancel" and he did.

Robin sorted it out with the fixer who I believe didn't get paid and wasn't used again. I asked Robin what the plan was.

"Don't worry, Mr K we'll sort it," he said, and he did.

A new fixer and a studio called Arco Studios were booked for the following Tuesday. Fortunately Neil was pragmatic about the situation and decided to return to London then fly back when the players and studio situation had been resolved. As we found ourselves with a day to spare, hoping the next sessions would be better, I suggested I hire a car and we drive to Innsbruck. I hired a Mercedes for the day and off we went. It was a good day so we decided to walk up a mountain. When we arrived at the top the view was breathtaking and to add to that was the sound of the cowbells sounding out around the animals' necks. I picked some small beautiful mountain flowers to take home and press them to mount in frames. I learnt much later that the flowers were a protected species and had I been caught it would have resulted in a fine! Oh dear! On arriving back in Munich I returned the Mercedes to the rental company, Robin found a good restaurant and we had a spectacular dinner with exceptionally good wine courtesy of Robin's generous hospitality.

The following day I visited Arco (they also recorded film scores) and was content with the studio acoustics and equipment and we were assured the orchestra would be good. Neil flew back, the orchestral playing was competent, the superb brass section, four trumpets, four trombones from the Munich Radio Orchestra and four horns, woodwind and strings from the Munich Symphony Orchestra. We only recorded Neil's compositions including several main pieces, five short scene-setter pieces, and six fanfares. The sessions went off without a hitch despite the fact the technicians spoke hardly any English and had certainly not seen the speed at which we worked! We mixed late into the night. Everything completed in an over twelve-hour day.

Ariola (now Named Löve Studios), Cologne 1969

This studio had recently changed hands, and the new owners were going to convert it to a "Pop" studio. However, for the time being the old staff were retained, as well as the equipment.

Boosey & Hawkes early June 1969

Eight hours recording, seven hours mixing.

Another successful two days' recording – a few of the composers from my hazy memory were Syd Dale, Malcolm Lockyer and Neil Richardson. Recordings released on Boosey and Hawkes' recorded music library, the Cavendish label.

Ariola, Cologne, KPM, June 1969

Main conductors and composers: Syd Dale, Johnny Pearson, Keith Mansfield, Dave Lindup and other composers' work conducted by Syd, Johnny, Keith and Dave. These recording and mix sessions were the longest period of time we had spent in Ariola. We also over-dubbed chorus on a number of titles by Syd Dale and Keith recording on the other two tracks on the Telefunken M10 recorder with the record amps switched to those tracks - the recorder had six playback amps - it was a strange set-up and worked well.

An enduring memory from those sessions was the recording of Johnny Pearson's *Sleepy Shores* which became the signature tune for TV series *Owen MD*. It was a chart hit in 1972 (released December 71) peaking at #8 and was in the top ten for three weeks, released on Larry Page's Penny Farthing label. I cannot remember the particular day I recorded the piece; it had an element of magic about it. John Fisher, the German translator, wanted to get a release in Germany – Robin refused. There is another story about the track that when the cut at 33⅓ rpm was sent to the TV company they played it back at 45 rpm thus it went on air at the higher speed so we had two versions. Johnny was invited to MIDEM (the music industry junket in January every year in Cannes) to promote the recording in France. It was a typical South of France sunny day. There was an upright piano placed at the edge of the water lapping the edge. The camera crew were filming Johnny for a TV programme. We made one take with Johnny playing and another requested. Johnny was about halfway through his performance when a moderate wave hit the shore and piano and on the receding wave the piano tipped slowly on its back with Johnny following it into the water, still playing! The piano was retrieved, along with a wet-trousered Johnny! The take was restarted as if nothing had

247

happened. John was not amused; all of us on the beach had a good laugh over the mishap.

One very well known composition by Neil Richardson was *Approaching Menace* better known as the theme for BBC televisions "Mastermind". This original recording was used from 1972. This was later re-recorded though not by me and in my opinion the new recording doesn't have the same impact as the original. These sessions were the second longest and most concentrated sessions ever at the end of which we were all very tired. Our musicians left at the end of the recording dates, leaving me, Robin and Graham for my mix sessions. As it turned out these June '69 were the last recordings we did with Robin at Ariola. Although I did not get involved I believe there was a falling out between Robin and Ferdy Klein. Löve was going to be refurbished as a pop studio.

Ariola, Cologne, Syd Dale, October 1969

Having set up Motive Music Productions with Syd, he told me he was going to have his own music library company and asked if I would produce for him. The library name was Amphonic Music Ltd.

These October sessions were the start of the library recordings. A selection of composers were James Clarke, Syd Dale, Brian Fahey, Bill Geldard, Max Harris, Don Lusher, Ernest Ponticelli and Dick Walter. It was quite a marathon. We spent 20 hours recording and 18½ hours mixing. This was the Yorkshire man who always wanted to maximise to get his pound of flesh from musicians and the engineer at the best possible cost!

Robin Philips was not pleased when he heard that Syd was setting up a potential rival to KPM and myself for the recording and production – unfortunately it strained their relationship.

Ariola, Cologne, Standard Music Library, September 1969

I had a telephone call at Lansdowne from Harry Rabinowitz, conductor and composer. I knew him from his days as Music Director for BBC Television Light Entertainment in 1960. He was now Head of Music (1969) for London Weekend television (LWT) and had heard I was recording in Germany for music libraries. He proceeded to ask me a number of questions about the musicians used, the studio and the fixer used to book the musicians. I was suspicious about his asking so many questions so was careful with my answers; I did not want to give too much away in order to protect Robin's interests. I told Harry we were using Ariola Studios in Cologne. He explained that LWT had founded a company called Standard Music to record library music, and he wanted me to organise a day recording and mixing in Germany. I booked the studio and put Harry in touch with the fixer – Ferdy Klein. I did not mention the use of our British rhythm section. It was important not to undermine Robin so I gave as little information as possible despite Harry's endless questions.

We flew out at some point during September for two, four-hour recording sessions with mixing to follow the next day. It was only a small orchestra with Harry conducting so we didn't anticipate any problems. Skip Humphries, the newly appointed head of Standard Music, came with us. As expected, it went off without a hitch, just the usual four track recording. Later, back the bar of the Königshof Hotel, Harry asked me how much he owed me. I told him and he then tried to pay me much more than was agreed, "because of the work I had done", and continued to pick my brains further. By now I was even more cautious with my answers as I knew that Harry always worked with Keith Grant of Olympic Studios located in Barnes London. After we flew back to the UK, I never heard from Harry again about recording in Germany. All future gigs went to Keith at

Olympic. Skip and I became friends though and he always called on me if there were any audio problems. In fact, Lansdowne was asked to install some audio equipment at the offices in Wembley for playing their albums.

WDR Studio, Cologne, 16th to 18th February 1971

The translator (and freelance upright bass player), John Fisher, on the KPM sessions asked me if I would like to go to the WDR studios to record the WDR Big Band for broadcast (minus the English rhythm section, using instead the WDR rhythm section) made up of the same guys we were using on the KPM and Syd Dale sessions. According to Fisher, the sound I was getting was better than anything achieved by the German engineers and he felt they could probably learn something from me. I recorded the band on 24th April in the afternoon WDR Broadcasting Centre, Wallrafplatz close to the cathedral. My recording raised a few eyebrows because it was hotter than they were used to. They had never seen an approach like it. They WDR engineers didn't appear to be happy the PPM was often in the red. It was recorded straight to stereo in a beautifully appointed large live studio – excellent acoustics. I enjoyed myself with the band, knowing the guys as I did by now and they trusted me to get the best for them.

AES Convention, Cologne, March 1971

As a member of The Audio Engineering Society I attended the 1st Central Europe AES Convention in Cologne, March 16-18 held at Esso Motor Hotel with a colleague Eddie Veale who had his own acoustic consultancy company, Eddie Veale Associates. It was a small gathering of like-minded audio people. Technical papers were presented along with an exhibition of 25 stands. Since then the convention moves around European exhibition halls and has grown to more than 100 booths. Conventions in the States were held once a year alternating between the East and West coasts, the format consisting of audio manufacturers' exhibition booths coupled with the presentation of papers on audio technology and recent developments in audio.

Katy Studios, Brussels, KPM, June 1972

This was the first time I recorded in Belgium. This eight-track studio was located in a suburb of Brussels called Ohain near Waterloo. The studio was next to a substantial villa set in a park with a large lake owned by pop singer Henri Markarian – stage name Marc Aryan. The equipment was adequate although I was not overly impressed, especially with the console by SAIT. The pianist on the dates was Francis Coppieters with whom I worked on some Cologne sessions when he played piano and Hammond organ – what a player! The rhythm section was the same as I recorded at Ariola with the exception of Les Hurdle on bass guitar. These were the only sessions we did at Katy.

I was due to fly ahead of the others on the afternoon of Sunday 18th June on a BEA flight. Instead I asked Robin to book me on a Saturday flight as I wished to go to the studio to familiarise myself with it and to set up. It saved my life because on Sunday 18th June flight BEA BE 548, a Trident aircraft, crashed just after take-off killing all 118 persons onboard. The others in the party, including Robin, fortunately were booked on an earlier Sunday flight.

Cornet Studios, Cologne, KPM July 1972

This studio located in Cologne Junkersdorf district – Aachener Strasse 1112a was a new well equipped sixteen-track studio with an MCI-JH-636 recording console, sixteen-track Lyrec recorder and Dolby "A" noise reduction. The studio installation was supervised and managed by an excellent sound balance engineer and producer called Wolfgang Hirschmann and owned by Heinz Geitz, songwriter, arranger and producer. This was the first studio I had worked in with transformer-balanced mains supply for all audio equipment and separate technical earth.

Our band was not large; we had the usual rhythm section Kenny Clare (drums), Alan Parker (acoustic/electric guitars), percussionist Jim Lawless and Francis Coppieters on piano. Brass sax's and woodwind were provided by the Boland/Edelhagen Band.

The sessions ran smoothly and the overall good feeling was helped by an excellent canteen with good quality home-made food – a benefit not lost on the musicians! I am sure the reader will appreciate with so many titles recorded I do not recall their names. Wolfgang and I became good friends over the years and worked together in London. Wolfgang was later appointed producer for WDR big band. Wolfgang was an engineer with good "musical ears". A tonmeister!

Cornet Studios, Cologne, Syd Dale, November 1972

Record and mix for Amphonic Music Library the usual Syd approach: get as much in the can as possible, which we did. It was a blur but enjoyable sessions recording to 16-track using Dolby "A" noise reduction. By now I felt comfortable with the whole studio set up and the equipment. Wolfgang Hirschmann and I became good friends our approach to recording was very similar and we were on the same technical wavelength. A most professional studio to work in.

Cornet Studio, Cologne, April 1973

I was invited by John Fischer to record a pop session with Ferdy Klein arranging/conducting. To be honest, it was a bit of a palaver. There was so much overdubbing: the rhythm section required many takes, then the brass overdubs and finally strings. It took forever to record take after take on laying down the basic rhythm track and overdubs even though the first takes were usable. The end result sounded clinical, lacking in the feeling one gets when recording all a band together. Now I knew why so many German recordings at that time lacked feeling. I was told that was the way they did it. It was good for studio income having to reset the studio for each of the section overdubs, especially good for my pocket as well! There was no interaction between the musicians and the resulting sound was clinical. But, the Germans were happy. I was not enamoured with it – it was not my way of working. The vocal was to be added later but not by me. All recordings sixteen-track Dolby "A" noise reduction.

Union Studios, Munich, January 1974.

Amphonic Music – Syd Dale

These sessions, continued Syd Dale's recorded music library at Union Studios, which was very well equipped with a CADAC quad desk and CADAC monitors. The guy running the studio was Mal Luker whom I knew from Morgan Studios London - a good engineer of the Rock 'N Roll "School". The CADAC console which was poorly maintained so I spent time getting some of the channels to work and changing indictor lamps in the monitoring section. Tony Campo (bass guitar). I previously worked with Tony in London he was now living and working in Munich, Vic Flick (guitars), Jim Lawless (percussion), Johnny Dean (drums), Francis Coppieters (piano). The rest of the band comprised brass and winds. Syd was conducting and myself producing.

It was a very cold windy January, we traipsed through the ice and snow walking to the studio slipping and sliding everywhere. Vic unfortunately was taken ill with what appeared to be flu – he didn't look at all well. The doctor was called and Vic (who was the guitarist on the Bond films from the very beginning) was ordered to take the pills and stay in bed for the duration much to his dismay. We didn't replace him and the sessions went ahead recording on 24-track with Dolby "A" noise reduction – 20 titles. We took the masters back to London and overdubbed Vic on the 20 titles in four hours. Vic was such a brilliant sight-reader that all the guitar parts were overdubbed at Lansdowne with no rehearsal – just a quick listen for tempo and then a take. I continued mixing on another day.

Decca Fonior Studios, Brussels, April 1974.

Sessions at Decca Fonior Studios were an eight-track recording. The monitoring system by CADAC was superb – very accurate. The large studio was the oldest in Brussels and had good acoustics: We recorded many items too numerous to list here. On this occasion Neil Richardson was conducting with Brian Bennett on drums – of The Shadows fame. Brian is a superb composer, he had written a piece called *Image* – a beautiful composition with lush strings and large orchestra. One everlasting memory was at the end of the take when the piece was in the can the whole orchestra clapped Neil assuming it was his composition, not so, Neil pointed to Brian and told the orchestra it was Brian's composition. It must have been weird and wonderful for Brian to hear the strings playing his composition while playing drums.. Another composition was by Johnny Pearson – *Heavy Action*, which was to become the title theme of *Monday Night Football* of the NFL (National Football League) America. This piece was also used for some years as the theme for UK superstars I was delighted with all the results and with a good team of technicians.

The British musicians on this date were Brian Bennett (drums), John Fiddy (bass guitar), Clive Hicks (guitars),Steve Gray (piano), and Jim Lawless (tuned percussion). The rest of the orchestra comprised a large string section – violins 1st and 2nd ,violas, celli, orchestral bass, woodwind, harp, four trumpets, four trombones and two horns.

Union Studios, Munich June 1974

These recordings (16-track) continued the work of building the catalogue for Syd Dale's Amphonic Music's recorded music library. Syd was always very well prepared and we recorded compositions by Bill Loose, Gerry Butler, Alec Gould, Steve Gray, Brian Fahey and Dick Walter. Some of these guys were not writing for KPM and Syd wanted to keep it different. Our London guys were the same line up as we had in January, with the exception of Vic Flick. The sessions were in the typical Syd style, a sweat in the control room reading scores, pushing the "knobs" and getting as much in the can as possible! At least we didn't work late hours into the evening – that made a change.

Morgan Studios, Brussels, KPM, early June 1975

English musicians: Brian Bennett (drums), Dave Richmond (bass guitar), Steve Gray (piano), Clive Hicks (guitar), Jim Lawless (tuned percussion), Duncan Lamont (saxophone), John Scott (reeds), Derek Watkins (trumpet).). Composers: Keith Mansfield – Johnny Pearson – John Scott – John Fiddy and Chris Gunning the latter two not present, their compositions conducted by Johnny and Keith. The other English musicians also contributed compositions to the sessions.

The band's line up was composed of trumpets, trombones, horns, woodwinds and a string section all drawn from La Monnaie (the Brussels Opera Orchestra). One composition still fresh in my memory is Keith Mansfield's *Grandstand Theme*. From overdubbing brass and reeds to 24-track from rhythm tracks that had been recorded in the KPM basement studios, which I transferred to Morgan's 24-track tape machine from the KPM ¼ inch – but one big snag did occur! The continental concert "A" was 444 Hz, the British concert "A" 440Hz so we had a tuning problem. The guys said the tape was running slow – a machine problem! Actually it wasn't. The answer was simple I used the tape machine varispeed to slightly speed up the tape speed to continental "A" 444Hz. Each of the composers present at that session conducted their own compositions for the overdubs. We recorded a number of titles with the large orchestra very successfully. Over the years a number of the KPM recordings became classic television and radio themes.

Some example are:

Grandstand – BBC's weekly sports programme. Composer Keith Mansfield.

Wimbledon – Theme tune to the Wimbledon Tennis Championships. Composed by Keith Mansfield.

World of Sport – ITV rival to Grandstand. Composed by Don Harper.

The Big Match – ITV sports channel. Composed by Keith Mansfield.

Gathering Crowds – Theme tune for America's Major League Baseball coverage. Composed by John Scott. *Heavy Action* – Intro to America's Monday Night Football programme. Composed by Johnny Pearson. *International Athletics* – ITV theme tune to athletics coverage. Composed by Keith Mansfield; Channel 9. *Cricket* – Original theme tune to Channel 9 World Series cricket in Australia. Composed by Brian Bennett; *US Open Tennis* – long running theme tune to US open series season. Composed by Keith Mansfield.

Authors note: These library music recordings were not generally available to the record-buying public.

Author: For the sake of completeness I have grouped all European library music sessions in this volume.

Chapter 13
Orchestral Recording Techniques – Microphone Placement

At the end of our trips to Germany, Munich and Brussels, I was asked many times about my recording techniques employed on the numerous sessions I recorded, to achieve the correct sounds on many of the musical genres, especially the clean hard-hitting rhythm sounds. It seemed easier to write a general paper on my approach, which is reproduced below. I wrote this paper in November 1975, 41 years ago. It was written with the non-professional in mind who had a close interest in recording techniques. It should therefore be remembered that at the time of writing we didn't have the technological audio processing tools, computers and hard drive recording we have today, and that recording and microphone types and techniques having further evolved over the intervening years, as has studio acoustic design. In Cologne and Munich I used all valve (tube) Capacitor (condenser) Neumann and Schoeps microphones. This Brussels studio in 1975 had an excellent collection of the latest capacitor Neumann's.

Making proper generalisation about microphone placement is clearly impossible, since it remains the most personal area of recording technique. This discussion, based loosely around a large "pop" orchestra studio session, indicates areas of doubt as well as the relatively few clear principles. Any session, even one as closely organised as this, is always crucially dependent on music involved. Below is the paper I wrote in

1975. A long time ago in terms of technology but many of the techniques and premises still hold true today.

Orchestra

My approach to recording was, and still is, science coupled with the art, good ears and an understanding of music in all its forms and the ability to paint a picture in sound with depth and perspective and spatial separation. It is also important to "get inside" the composer's head, to fully understand what he wants and what he hears in his head when writing a score, and transpose that from the studio floor to the recording medium. It is equally important that composers have a good understanding of recording techniques. In a studio recording set-up/layout, don't separate musicians from the playing positions they are used to in an orchestra. You won't get a good performance if the player feels uncomfortable. And *very* importantly, know how to mic up an instrument to capture the whole sound. I have seen engineers stick a mic down the bell of a sax then wonder why it is a raspy rotten sound. The sound comes from the whole of the instrument – learn where sounds emanate from on all instruments of an orchestra, a simple learning curve and the result will be a better-recorded sound. One can normally only generalise when discussing microphone techniques, as it is perhaps the most personalised area of the art of recording. So, there are several precautions that should be taken in writing a paper about such techniques. During this paper I will make reference to families of instruments and sections of the orchestra, while not necessarily tying things down any specific microphone type other than a dynamic or a capacitor. It is also worth bearing in mind that most sound engineers have pet types of microphones for their particular type of recording work and choose the correct types for the style of recording taking place. I will be discussing the "Pop" approach to recording, which takes place under fairly dry acoustic conditions, such

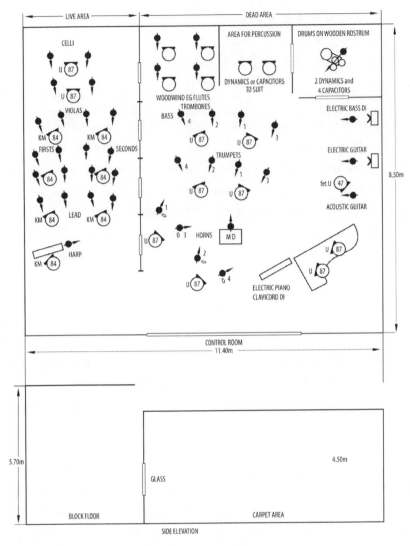

Figure 1. Studio Layout

as a studio having a reverberation time of around 0.4s at 500 Hz and reasonably flat in reverberation time over the range 20 Hz to 20 kHz (depending on the design of the studio acoustics). A looser approach would be a studio reverberation time of 1.5 to 2 seconds, rather than the classical approach, which requires a larger space and a longer re-

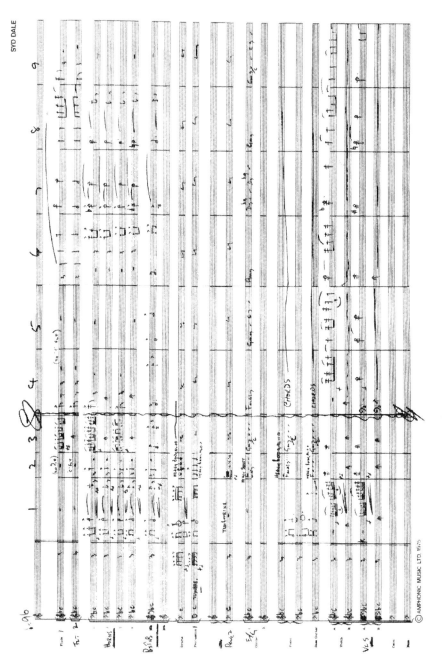

Figure 2. Pop overture to bar

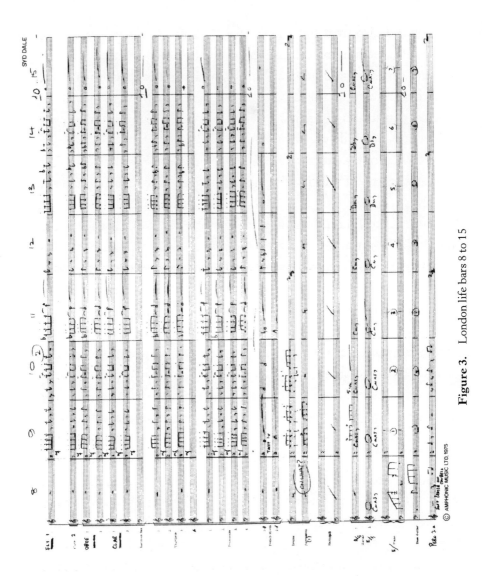

Figure 3. London life bars 8 to 15

verberation time. Musicians generally dislike studios that have short reverb times – it makes them feel uncomfortable. The control room acoustic has to be well designed for accurate monitoring.

When dealing with musical instruments (I discuss a medium sized orchestra) we should really consider the instruments section by section but also treat them overall as a whole. Obviously, they are related by section and by ensemble to the musical score. You can discuss certain microphone techniques for specific instruments but you must remember that any technique applied always relates to the type of music and the particular type of sound that you are aiming to achieve. So while we can give examples of how to mic up specific top sections of the orchestra, that is any section of the orchestra that is going on top of the rhythm section, variations on this theme will naturally occur.

Specific techniques for dealing with the rhythm section have evolved over a number of years and there are reasonable generalisations that can be made about how to record a rhythm section correctly. Engineers never usually mention these value judgments because they are normally working in conjunction with a producer, who may not have a clear knowledge of what is required. It has always been my belief (and I'm thinking of many other engineers) that having recorded a good rhythm section sound with a good gutsy, punchy sound in context with the given type of music, you can usually sit any other family of instruments on top and make the thing happen. I have found this to be the case whether dealing with large pop orchestral works with strings and brass, or just coming down to the middle of the road recording or, indeed, a heavy rock piece with strings, brass, woodwind and heavy percussion etc. and other synthesised (the very early days of synthesizers) electronics that go on top, usually, it does work.

It may look as if many outs have been taken during the course of this introduction, in avoidance of definitions, but as the experienced engineer knows it is extremely personalised. Being asked to describe a microphone technique is really rather like a painter being asked how and why he painted a particular picture and why did he choose those paint colours. It's very much the same thing to ask an engineer who is acting as the artist in this sense. He is painting pictures with sound for depth and perspective. How does he begin to paint them and where does he begin? The question "why?" without qualification is impossible to answer by painter or engineer. The painting (recording) is an art form. So I reiterate that what follows is a fairly generalised, although often specific, example of how to treat a fairly large orchestra. Later on we will deal with vocal overs, for completeness' sake, but this is a whole other subject in itself.

I've always found it an obligatory practice to discuss with arranger and producer the type of sound that we're going for on the date (session). It is absolutely essential, when dealing with either studio set-up or studio recording acoustics, that the precise nature of the music be known and the purpose for which it is being recorded. Do the desk preparation homework first; it makes life a lot easier on the actual session.

We must also consider the question of separation, which some engineers either ignore or fail to understand sympathetically: some use too much and some get it right. The right degree of separation for the particular type of music is very important to the overall feel of the final result. Even in a moderately acoustically dead studio good use of no separation can be made to suggest a feeling of depth (air around the sound), width and perspective. Very often it can be overdone, though, resulting in the instruments of the orchestra sounding like they've been completely separated out, even as if they've been overdubbed to a clinical degree so that when the whole thing is mixed it doesn't come together as a whole and can sound stilted.

When people say "Well it's a great sound, but it doesn't *feel* right", usually it's because of bad technique in any of the miking, use of studio acoustics, separation or the scoring. The final sounds on any recording only come together with extremely careful placement of the microphones – even moving a single microphone within a section by a few centimetres in any direction can make a complete difference to the overall sound of that section. A very good specific example is that of a large trumpet section, given later.

This paper takes the layout of an idealised medium to large "pop" orchestral set-up. We also have an accompanying string section recorded simultaneously. For points of discussion, it will be easier to refer to **figure 1** which shows a typical middle-sized layout in a medium-size studio (for the sake of clarity cans (headphones) are not shown). Apart from all balance engineering thoughts, the first concern is that the musicians can see each other and the MD and in turn the MD can see the whole orchestra. Good communication is key. There's nothing worse than an engineer who does not consider the comfort of the musician or MD. As strange as it may seem, comfortable musicians play better because they can react to each other.

The line up in **figure 1** includes: drums, percussion (for this example I've put in vibes, the toys, the Latin-America family of instruments, conga drums (tumbas), wood block, tubaphone, marimba and timpani), electric bass, electric guitar; acoustic guitar and piano doubling electric piano doubling clavinet. Woodwind is four players doubling; cost being a very important consideration in commercial recording. The four woodwind: two flutes doubling piccolo doubling tenor sax and bass clarinet. Two oboes doubling cor anglais and contra bassoon. The brass section consists of four trumpets, four trombones and four horns. Strings are 12 violins (ten firsts, six seconds), four violas, four celli, and harp. Some would say this is not an internally balanced string section – it all depends on the composing. Incidentally, for sheer weight, my personal preference here would be for sixteen violins ten firsts and six seconds, four violas, four celli and two orchestral basses. The accompanying score, **figure 2**, shows four woodwind, four horns, bass, trombone, drums, percussion, electric guitar and the string section, and is shown purely as an example of the writing; the other score **figure 3**, a smaller ensemble although the orchestras are not the same in numbers, it is useful for examples later.

The orchestra in **figure 2** represents a version of the session we are discussing not to be confused with a sample studio set-up in **figure 1**. The type of writing typical here is heavy boogaloo rhythm section, really punchy brass on top, with string backing punctuated by flutes. Flutes are written either with the horn section or as an independent section between the strings and brass; the woodwind work with horns in bars 2 and 3 – note the horn voicing. In this particular example we're using only two woodwind and not the four, as of course it very often happens in writing that we do not use all the instruments all the time otherwise we'd live in a very boring world!

In **figure 1** we have a good situation: the MD can see everybody and they can see the end of the stick. From the engineering aspect, it's comfortably laid out to give good separation – instruments that give the least amount of output, woodwind and string sections, are both well away from the brass and well separated, particularly the strings, which are divided by the screens. As indicated, in this studio these screens have small windows in them set at an angle, usually downwards, to reduce audible reflections. Woodwind is close against the wall and are screened either side. The screen between the percussion and the woodwind is fairly heavy and reaches from floor to almost ceiling height. The drummer is enclosed by two screens, with a gap between, and once again he has small glass panels to be able to see what's going on.

Bass, electric guitar, acoustic guitar and keyboard sections are fairly well spread out and this can present a problem in an orchestra of this size with regard to feeling and interaction. However, the piano player can see the acoustic guitar, the electric guitar, the bass player and the drummer and vice versa, and the three electric instruments are sitting next to each other, divided by two screens approximately 1.5m high. It's quite normal in this type of set-up to give percussion, drums, bass – in fact all the rhythm section – headphones with foldback, various mixes for the individual players or the total orchestra. Separate foldback for the rest of the orchestra from a different foldback circuit. Also, the MD will be in direct communication with the box via cans. In many studios the MD will be able to speak directly to the individual player foldback. This makes it easier to talk in confidence to any player over some delicate point without public embarrassment. The MD might also have PA to talk to the studio, as well as to talk to the control room thus avoiding having to shout over the whole of the orchestra – players do have tendency to talk among themselves in between takes.

For a smooth session it is important that all these factors, and others relating to space and the comfort of the players are right, particularly when they have to sit on a studio floor for three hours or more. Everybody must be absolutely happy and comfortable in order to achieve good results. It can be argued, of course, that this is an unusually large session and with today's (mid 70's) multi-track recording techniques we would not record all the strings and all the brass together. It is a fair point. In fact we would usually record just the rhythm section overlaying brass, woodwind and strings individually. However, my experience tells me time and time again that the results are so much better when the whole orchestra plays together, especially with a large brass section, because the rhythm section is able to hear what is being played and the drummer can drive the brass section – and giving it some expression, something that cannot be achieved on overdubs, which lead to a no-feel result, aa clinical performance. Brass players love to get a large rhythm section behind them to get them working.

The next problem that arises, of course, is the enormous number of channels that we're going to use to record the orchestra. Maybe the console is limited in its number of input channels; compromise can be difficult but technically not insurmountable and not relevant in studios of the present. Then we hit the headaches of the actual mic techniques needed to achieve the separation necessary to get the close sounds, to use the room acoustics to their best ability and to use distance between player and the microphone in conjunction with whatever limited room acoustic is available. The string area here is wood block floor, which is relatively liver than the adjacent carpeted area. The ceiling height also differs between the two. You would expect a reverberation time in the carpeted area to be about 0.3s at 500 Hz, with the bright area to be getting into 0.8 or 0.9s. Now we turn to the mic technique itself, section by section. The orchestra divides naturally into the following 1) rhythm, 2) brass, 3) woodwind, 4) strings, including harp.

Rhythm

Most studios have their own preference for their rhythm set-up with the rest of the orchestra set around this. In my extensive experience, hard and fast rules don't always apply. Today's (mid '70s) medium-sized studio acoustic thinking tends to be towards mixed areas for the rhythm section. The basis for the whole of the recording is the rhythm section and, with the drums, it seems appropriate to give specific favourite mics. These are based on personal preference – other engineers use different types with good results. As I have said, it is not possible to generalise but it is possible to personalise.

Bearing in mind that we're given a rock rhythm section, my own particular preference for this type of drum recording would be as follows: bass drum: dynamic mic, eg. AKG D12. Snare drum: two microphones – one on top and one on the underside. At this stage it must be mentioned that we would always use equalisation carefully to achieve appropriate results for the type of music that we are recording. The top side of the snare has a Sennheiser MD441U, with a Neumann KM84 on the underside. The KM84 capacitor mic is used very close to the underside of the drum, so the 10 dB head cut should be employed and the mic channel should be phase reversed – under snare sound is 180° out from the top of the snare. Hi hat: KM84. Small tom-tom: AKG D202. Floor tom-tom: AKG D202. On the overall kit: Neumann U47s.

It is difficult to discuss in detail the exact positioning of these microphones as it very much depends on the player and the type of kit employed however, we can outline a few basics. Most drummers in studios now employ the bass drum with the head removed, and therefore preference would be for the mic to be close inside the drum shell close to the back of the front skin. The closer you get to the centre the more you get of the actual impact: the farther you come out from the drum, the more you get the "overtone", and more of the harmonic structure of the note rather than the basic fundamental. I can only stress once again that final mic placement can only be made by listening and by knowing the drummer and the kit. The snare microphone is normally placed on the edge of the drum itself, pointing in towards the skin. This is more for convenience than anything else because if it's too obtrusive it gets in the way for riding on to cymbals and on to tom-toms, particularly the small tom. There is no problem at all with the underside mic placed right underneath the drum, slightly outside of the snares themselves. The question arises once again: what sort of sound are we going to get? And again, this is where the ears come in. When mixing a snare sound, equalisation is all important. You must be careful and have the right sort of feel when putting the two sounds together. Without actually playing an example of the different sounds that can be obtained by this technique, it's necessary to leave it to the intelligence and imagination of engineers. The tracking and eventual mixing of the orchestra as it affects mic placement will be discussed later. The Hi hat presents no problem, with normally a capacitor microphone, or indeed a dynamic if one is available, placed over the hi hat slanting toward the perimeter. Small tom tom mic is placed at an angle downwards and towards the edge of the drum. The same applies to the floor tom. The bass drum (kick) was miked internally with the front drum-head removed – on big orchestral work the recorded sound works well by picking up the kick on the overheads. Overall kit sound comes from the two microphones, two U47 on top of the kit. These must be used with the greatest discretion; once again the artistry of the engineer comes in and it is difficult to define just how much of the overall sound one would use without listening to the drums with the rest of the orchestra. So many producers and engineers fall into the trap of listening to individual instruments, equalising them, then throwing the total sound together later. Often it doesn't sound right. Final equalisation, final mixing, whether it be monitor mix or final reductions (mixdown today), must be anticipated by listening to the overall orchestra. Even a few millimetres' movement on the fader, two or three dB difference in hi-hat equalization or minimal variation of the overall kit's sound contribution can make an enormous difference to the overall sound obtained.

Finally we should mention double kits and other variations. Obviously use extra microphones as necessary or perhaps just one between two tom-toms together if dealing with two small and two large tom-toms. Ideally it would be better to mic them separately, but this depends on the number of channels available. With one mic on two tom-toms it is better to mic above and directly in between the two adjusting the sound that one hears.

For me, there are three methods of "miking" on the bass guitar. One is direct from the pick-up of the instrument – not always successful because it depends on the player and instrument. The second is a conventional microphone pick-up from the speaker output and the third from the direct output of the loudspeaker coil. I have known many engineers to use a combination of all three and this sometimes works. However, my own personal preference has been to take a direct output from the guitar pick-up and inject this directly into the desk – of course with eq and the other et ceteras that go with it. This can usually result in a hard, dry, tight, forward sound. When using combinations of mic, direct pick-up and direct from the speaker, very careful balancing and equalization of the three channels is needed. In many cases, it results in over miking and the simpler method of one of the three usually works better. For best results you have to know the player, the amplifier and the instrument concerned. My particular instrument preference in recording comes down to Fender Jazz or Precision and/or the Rickenbacker. This, once again, depends very much on the type of music. All three, though, get a good bass sound with individual variations for the type of material that requires it. This brings up the question what type of material requires a Rickenbacker sound? This is a difficult one to answer because it is a judgment call by the player, producer and engineer whether this type of sound produced on the session is appropriate. I should also mention phasing: if using direct from pick-up, microphone and direct from output it is extremely important to be sure that all signals are in phase. Most consoles can cope with phase reversal.

It's always best to put the bass player as tightly and as closely together with the whole session rhythm section as possible, consistent with requirements of separation. Even though they might be wearing cans and can hear each other, they must have this intimate communication and the feeling of tightness within the section, which can only be achieved by close-knit work. Many engineers, because of separation problems and for other fads best known to them, tend to separate these players into little boxes. In my experience this leads to a musical disaster. The players feel the same way. Separation between bass and drum depends on music and bass volume. It is usual to get 20 dB of separation, even when they're close, between bass and drums, and with careful use of high pass filtering there is usually no problem in separating bass and drums or any of the other rhythm instruments. In fact, drums-to-bass is generally better than 30dB with miking and bass to drum better than 25 dB. Normally this is achieved without any problems at all. On recording the bass, I prefer compression to smooth out the unevenness of the instrument or the playing of the notes, usually employing a ratio between 2:1 and 4:1 and a compression factor of not more than 4 to 6 dB on the overall spectrum of the instrument, thus working towards a hard, tight, forward sound. I do not believe in limiting or compressing any of the other instruments as I record, preferring instead to get them down as they are with their dynamics unless production reasons dictate otherwise. Electric guitar is usually not much of a problem. A good dynamic or condenser rule will suffice, placed against the amplifier. Most of the effects are being created by the player himself, with others done later on by mixing. From a separation point of view acoustic guitar can present a bit of a problem in this layout. If the brass were playing loud, an amountwould spill into the acoustic mic together with a lot of electric guitar, due to their closeness. However, with something like a Fet U47 type, placed closely on the instrument, separation can be achieved. In the light of this set-up I'd suggest using just one mic and not two as many people do placed where the strings go across the rose hole (not too close) the other at the back front pointing toward the wood not in the f hole. It is very much a matter of personal choice and type of instrument being played.

With this sort of line up the acoustic guitar probably isn't too prominent, just clanking away, so we use a single mic approach. If the acoustic guitar were used here as a front instrument then the recording objective would change, suggesting a stereo treatment. If we were overdubbing then the approach would be entirely different once again. With acoustic guitar in particular, the acoustics in the

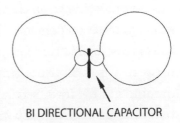

BI DIRECTIONAL CAPACITOR

Figure 4. Congas

studio and the microphone distance from the instrument would be governed very much by the type of sound and the type of equalization employed. It is very difficult to be more specific than that. In this particular instance with a Jumbo guitar, the miking would be on the sound hole of the instrument itself and placed sympathetically to achieve the required sound. Use your ears!

In this particular set-up, it is quite likely that the piano will be recorded in stereo but not perhaps on every number. I would use two capacitor microphones U87 type. The KM84 and U87 were my workhorse mics in this studio (I would have preferred the valve Neumann's but they were not in evidence in this Brussels studio). One would be placed fairly high up on the strings, the other placed where the strings cross. At this point I should really define the piano as a Steinway or a grand piano of similar nature. It is once again, and it seems like another get-out, very difficult to define the exact position of the miking on the instrument itself. One can generalize, as previously, but in practice the movement of both microphones, even by a few centimetres, can make an enormous difference to the sound. Pianos in heavy rhythm section set-ups have tended to be difficult to record because of the amount of bass that will enter the mics from the bass and bass drum. In many studios this leakage can run down the walls and, if the piano is close to the walls, into the piano. Therefore piano traps are employed (70s) as a preventative measure and to give good separation. One can to a certain extent alleviate such bass problems by careful use of high pass filters, but this defeats recording a clean piano with its full frequency spectrum (if the production demands this approach). The piano can be classed as a percussion instrument and has an enormous range. If we're going to use it to its fullest extent in any orchestra, whether it's pop, classical or middle of the road, it is my feeling that we must record the widest range possible consistent with separation. We'd use the right amount of equalisation to get the right amount of poke out of the sound.

273

Although not shown, jangle or tack piano must be mentioned – the upright piano with the tubby sound, the piano that either has drawing pins in the heads or that has tabs and foot pedal arranged to give us the jangly sound. There's no mystery about recording it, being simply a question of poking an ear in the back, listening to the type of pattern that's being played and placing a microphone on it. Electric piano is very much easier to record than its acoustic forbears. My studio preference is for the Fender *Stereo 88* piano (very popular in that era) either using two microphones for stereo effect and spreading it across two tracks, or direct injecting still in stereo via two transformers into the desk. Direct injection does alleviate any possible separation problem and gets a very tight, up-front piano sound. This is probably many other engineers' favourite way of recording it as well. Clavinet is, again, direct injected, giving a very forward, tight sound. There is plenty of control with the eq and no worries over separation. Although not shown, I would mention in passing the organ. It may be C3 or whatever, but the Leslie is of course the key to the whole sound. Preference here is for a double miking technique on the Leslie: one mic on the top and one right on the bottom of the instrument's Leslie speaker close to the cabinet. It's better miked at the back of the cabinet where one can get directly into the cabinet itself.

Figure 5. Trumpet Section

274

Percussion

Percussion is a difficult section in many ways to cover because of the vast number of instruments that the engineer has to handle. It's really a subject in itself but I would like to give one or two examples. Congas have three ways of miking, possibly four ways if we look at it in stereo format. Two mics go on top, one over each of the drums at about 0.5 distance, or at least a sufficient distance to give the player room to raise his hands above the instrument. This gives us a stereo picture. The second possibility is to mic from underneath and inside with dynamic microphones; I have tackled it with dynamic microphones and had excellent results. A third method would be to use one microphone over the instrument itself, of course using equalisation together with careful listening. Lastly one bi-directional capacitor may be placed edge-wise on between the Congas . Separation this way is excellent (**figure 4**). Bongos could be treated in the same way although it's very much more difficult and restricting for the player. My own preference is to mic the bongos either with two mics if the music calls for it – from the top side pointing directly out to the skins, or with one microphone in between the two drums themselves. Tambourine is an interesting instrument to mention here, not only for the type of sound that can be achieved by equalisation but also by the differences in sound the player can get either using it with the skin removed or with skin on. Once again you have to listen to the sound and determine the exact placement. One of those instruments that's treated lightly by many but its rise tune is extremely fast. Timpani put the frighteners on many engineers despite the fact they are not really that difficult to record. Depending on how many timps are being used, I would use one mic between two, or two mics between four, spaced at 1m above the instrument. Usually a capacitor microphone is used, adjusted in height and position according to the type of sound required. They're very versatile instruments. A particular incident was on a film where we had to recreate the sound of very high flying turbo-prop aircraft with associated low rumble; this was achieved by using the large orchestral bass drum beaten very softly; mic at 1 cm and equalized. That example also gives some indication of the tremendous sensitivity demanded – the placement of microphones on an instrument to achieve a given sound is often only arrived at after much trial and error.

Brass

There are 12 brass instruments in this set-up, a large complement. Bearing in mind the type of music and the acoustics of the studio my preference for miking would be to divide as follows: the four trumpets would be handled with two mics, one between each pair. Trombones would also be miked one between each pair. Ideally the horns, if there are enough channels available, should be miked individually for this type of recorded music, but in practice you might use one mic in between each pair of players. It sounds easy but in practice is not quite so. It is of paramount importance that the mics on all these instruments (and I've no actual need to say this to the hardened professionals among us) have to be placed absolutely correctly to achieve the correct *section* balance. In bars 9 to 14 on the example shown in figure 3, you will see how the sections are working together and the relationship between the woodwind, trumpets, trombones and horns. Musical readers will see how important it is that the balance be absolutely right. The critics among us, and I include myself in this, will say,

"Well yes it's very easy to describe mic positions but what happens with distance? What about overload at close proximity to the brass?"

The answer is very simple. Preference on this particular set-up, and indeed in general, would be to use the U87, with 10dB head cuts. It is again difficult to say at what distance to work from the instrument, it's a question of logical thinking combined with a

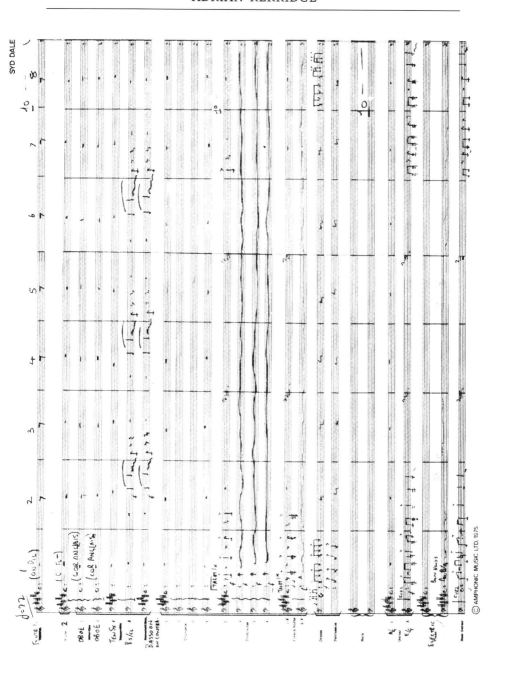

Figure 6. Juggernaut to bar 8

good feeling for what you're recording, but one could put an average distance between, for example, trombones of 1m or 1.5m from the instruments.

Room acoustics play an important part here. The same with trumpets, but then of course we have further problems with open and closed brass – my own particular preference is to ensure that when the instruments are opened and closed I ask the players to cooperate by moving in and out accordingly. The section is probably divided into two and two; if the exact spacing between the players and the mics is not spot on so that the perceived strength of one player is greater than that of the other, then the result in the box can be disastrous. In figure 1 I the trumpets are set I and 3 and 2 and 4, their normal positions in the orchestra. Frequently the leads change though, depending on the musical requirements.

Now, some engineers would prefer to get it as in **figure 5**, which obviously makes life very much easier. With these two microphones brought up on two channels there is quite a lot of leeway for error, but it's very important to listen very carefully to get a good four-part harmony especially should they be playing *divisi*. **Don't lose the lead trumpet in the balance**. Other complications can arise since trumpets to the section can be further divided into a mixture of flugelhorns and trumpets. This may vary to the extent that the trumpets may, for musical reasons, have mutes in. If this is the case then it's a question yet again of listening and adjusting the balance to ear.

Out of all the brass family the horns, because of the awkwardness of the instrument, are the most difficult to mic. Current thinking is to mic the instruments directly from the back but a few years ago (and for some engineers today) the preference was to use a reflective screen behind the instruments and to mic the sound on its reflection. I don't disagree with the old ideas but I don't think it makes a particularly good sound for this style of recording. My own preference is to mic the instrument directly on the bell, once again at a distance of 0.5m to 1.5m depending on the acoustics and playing. There is one snag here, however: horn players spend many years losing all the watery, raspy sounds that you hear when you put your ear very close to the instrument. With careful placing of the mic this can be avoided and we can get an extremely good tight, close horn sound without a raspiness that I've heard on too many recordings. In the example that's shown in fig. 3 – since the horns are playing long notes, this is not a problem. Separation between trumpets, trombones and the horns in the example given is not a problem because of the proximity of the instrument mic itself. One question that immediately arises is that of separation between the brass section and the acoustic guitar. You may run into problems if a reasonably omnidirectional mic such as the *KM53* type is employed, but if a mic is used such as the one we discussed, the fet U 47, fet 87 *(field-effect transistor)* or a similar cardioid type, then really with close miking, this should not be an issue.

Woodwind

It will be seen in the example given in **figure 6** that the woodwind are working independently from the brass section. My own miking preference for woodwind in this case of two C flutes is to place the microphone fairly close to the lip plate but away from the direct breath of the player. Distance would be approximately 10 cm from the lip plate itself and slightly to the right as one looks at the player. Many of the purists will say this is far too close to mic for a natural sound. I agree but for the given example this would be the right thing to do as it would give us good separation and a very tight sound once again. If we now talk about a more orchestral sound where a more fluent, transparent sound is expected then the mic technique and approach would be completely different. The microphone might be placed 2m from the player, possibly at a 45-degree angle and back. On overdubbing this would not be amiss.

Thus, when you talk about miking the woodwind family it is even more essential to have a very intimate knowledge of where the sound comes from and the type of music providing the context. Another good example here is of the bass flute, an instrument that I personally love and I feel is not used enough. Is beautiful when well recorded, but gives small output indeed. So the best results for me have always been achieved by miking very closely on the lip plate, again slightly to your right of the player. As with the C flute, we do not get the breath of the player but emphasise the beautiful, sonorous timbre that the instrument possesses. The alto flute is a similar example, although its relatively low output is not as low as the bass. Piccolo speaks for itself.

The oboe is a very resonant instrument and gives a good output. I would normally mic it at about 0.5m in front, with the mic face looking down on the instrument itself i.e. the cardioid side of the mic looking down. It's necessary to avoid the break in the instrument and to mic towards the top end because any difference in output will tend to be at the break itself, which is always difficult for the player to smooth over. Cor anglais is an extension of the oboe, miked in a similar position as shown in **figure 7**. Certainly my own recommendation for miking this is not closer than maybe 0.75m, although it's hard to tie down precisely. Clarinets yield to similar miking technique to cor anglais and oboe, once again depending on type of music, type of material, type of sound. For pop close work maybe 0.75m is about right; for larger orchestral work, it's a question of taking the mic away and listening within the orchestral context. Bass clarinet is another favourite instrument of mine. If it's in the pop context this can be tackled with two mics because the instrument gives out

Figure 7.   Cor Anglais

part of its range from the body of the instrument and the lower notes from the bell. One microphone about 0.5m from the instrument, spaced equally between the bell and the main body, will also suffice. The contrabass clarinet produces, when used in the lower register, sound from the bell. When very closely miked and well equalised it can make some fantastic sounds that are completely unrecognisable as the instrument yet adds tremendously to the overall sound of the mix.

The bassoon should, in the pop context, be miked fairly closely. From the engineering point of view it is, as usual, a question of sticking your ear around it. Although it sounds crude and simple, it is also very effective in determining exactly the position of the sound and therefore where to place the microphones. It has been my experience that, when used in pop writing, arrangers tend to stick to one register. Such sound will come out at a particular point of the instrument so put the mic in close proximity to the area of the instrument that is giving the best sound. Once again, it's very hard one to define and a question of using your head. The contrabassoon sounds an octave below the bassoon. It is a difficult member of the woodwind family to attack because it depends on the writing. It is also very large. A specific example is shown in **figure 6**, bars 2-7, where it is treated as an interesting sound on its own. It would therefore be very tightly miked on the bell of the instrument. It is also interesting to note that it is in unison with the bass clarinet, sounding an octave below, but the relation of the two sounds would depend on musical intention.

Saxophones cover the whole range: soprano, tenor, alto, baritone and bass. Most commonly used are soprano, tenor, alto and baritone: bass saxophone is used occasionally but a typical section in a band, a large band, would be two altos, two tenors and a baritone. Once again it's very difficult to tie down actual mic distance but it would be pretty close and in the region of 0.5m to 1m from the instruments.

Normally, you would head for in between the bell and the body of the instrument. The sound, as we're all aware, comes out from mostly down the whole length, except with a soprano saxophone where we get the sound largely from the body. However, I have had instances of recording with the soprano saxophone where I've decided to use two mics, one on the body and one placed at the mouth of the bell. There should be a qualification that saxophones play many different roles in the orchestra, playing a particularly wide range of different types of music and therefore it is very difficult to be absolutely precise on the miking technique. It is a question once again of the sensitivity and the feeling of the engineer and his musical ability to interpret what has been written. In many rock sessions for example, particularly where we're concerned with the baritone saxophone, it's acceptable to stick a mic down the bell of the instrument itself and equalize the hell out of it, resulting in some remarkably good sounds. The same thoughts could apply to a lesser degree to the brass section. Here, especially, the engineer must be aware of the total sound pressure level the mic will accept; many of the mics that we use today of the capacitor types do have attenuation on the head to prevent overloading close to high output instruments.

Strings

In this particular instance we're talking about 12 violins, four violas and four celli. The context is pop writing. Violins are divided into six first and six seconds; violas are written in four parts and the celli are also in four. Preference for technique here is to mic the firsts with two microphones of capacitor types such as KM84 and the

Figure 8. Violin Pair

seconds likewise, so we have four microphones covering the whole violin section. The front mic on the first is placed at approximately 1.25m above and slightly to the front of the player, looking directly towards the pair of instruments. This can be seen in figure 8 and the mic position relating to the overall studio also by looking at **figure 1**. As there are four players to each mic on the back set of violins these were miked at a distance of some 1.50m and at a slightly different angle to pick up the four players. It will be appreciated, however, that mic placement is extremely critical and it is down to the experience of the engineer to place and to listen. The violas are given one mic between two, placed similarly to that of the front row of the first section. Celli sometimes present a problem in the studio, but for the purposes of this type of recording I decided to split them down into two desks using two U87s and positioned as shown in **figure 9**, This whole technique employed on the string section is open to discussion by engineers for it is very much tied up with the room acoustics, the type of music that's being recorded and the other caveats that go with it. However, we have to remember that in this type of layout we are also fighting a separation problem, to which I will refer later.

The harp probably gives most trouble from a separation point of view, discounting any equalisation or high pass filtering. My personal preference (and I'm now talking about total live recording, where one needs good harp separation) is to take a small dimensional capacitor microphone such as a KM54, bound round with some form of anti-shock mounting such as sponge rubber. This goes in the second hole down on the back of the instrument facing the soundboard **figure 10**. The purists amongst us will shudder but in actual practice with the right type of eq the sound can be phenomenal – very close and truthful. The other placement is just off the soundboard. This, though, has problems because the player pulls the harp back into the body while playing. Many harpists get a very easy time on sessions, because they play six bars for the intro, rest for another 84 bars and then play another two bars. So you can have a problem if you are placing the mic on the soundboard, for the player cannot rest the instrument without moving the mic. A third system is to place the microphone just off the strings, fairly well towards the top end and to the right of the instrument as indicated in fig. 10, position 3. The only problem with this approach is that although you will get a very good sound the mic has to be placed diagonally and fairly close and can become an obstacle for overenthusiastic harp players, particularly on runs. The problem that also arises here is picking up another instrument playing nearby. With close miking the output of the harp being fairly heavy so this doesn't normally interfere. The dictates of studio set-ups seen in **figure 1**. It doesn't always give ideal conditions with many instruments. A point that immediately comes to mind is the proximity of the harp to the front desk of string players not a good situation from a separation point of view.

Vocals

A typical vocal group might be four boys, three girls. My preference for miking these might be as indicated in **figure 11**, the vocal group in separation room or booths of sufficient proportions to ac-

Figure 9. Cello

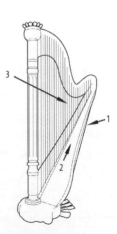

Figure 10. Harp

commodate them comfortably with the right degree of air conditioning. There must be good visual contact with the studio, and good foldback facilities. Some vocal groups prefer to work on cans, some prefer to work to speakers. Whichever it is, it's important to realize that vocalists like to be very comfortable and feel that the room they are in is working environmentally with them and not against them. In **figure 11** they are all facing towards the studio looking through the glass and quite separated, This sort of set-up would suffice only for certain types of writing. If the writing is close harmony where the group need to have a strong musical feel with each other and get working together, then the set-up in **figure 12** would be better. It could be argued that as they're isolated it doesn't matter where they are but this point of view is short sighted as they need to have communication with the MD particularly if dealing with *colla voce* passages. They retain some form of communication with the studio. It is a bit of a compromise but remember they have foldback. You have a much better chance of a very good internal balance with the group. You will see in the diagram that two small loudspeakers are placed facing the dead side of the mic, one for each of the vocal sections. I normally never record the vocal group using pop shields, preferring groups to have good microphone technique themselves with diction for the letters p, b and d well under control. The same applies to solo vocalists.

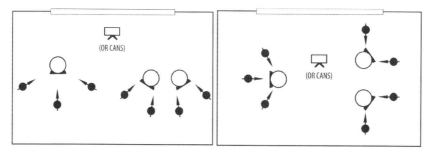

Figure 11.   Figure 12.

285

Many circumstances force us to use pop shields, although I know of many solo artists with extremely good mic technique who can, and do, work very close to the microphone.

Separation and Scoring

We should discuss the requirements of separation. These depend on the heaviness of the rhythm section compared with that of the strings: the heaviness of the brass section compared with that of the strings and the heaviness of the brass section and the rhythm section compared with the woodwind. Now, as we can see from the set-up, the strings are reasonably well screened, because in the studio the screens go front, floor to ceiling for almost the length of the studio. The flutes are screened. Separation difficulties are going to occur between the brass and the strings. On this particular session it was not untypical to get separation from *ff* ("fortissimo" and meaning "very loud") *brass* to strings of 10dB. Brass to flutes came out in excess of 15 dB. Rhythm section to string section was no problem whatsoever – certainly in excess of 15 dB. Percussion to brass or rhythm to brass was also no issue because of the close miking on the brass section and the screening of the heavy output instruments such as the drums: separation was probably around 18 to 20dB. Thus, it can be seen that there is enough flexibility of separation for mix down. The major problem with really heavy rhythm and brass will be as against the strings. For this very reason it is common practice to overdub a string section to maintain a good degree of separation and to provide for good control later on in the mix down. Brass can also get into the piano. The piano lid is generally raised on the short stick and covered with some material – the musicians' pet hate, because the piano sounds like a dead lump of wood and doesn't sound for them. From the engineering point at view, this will give good separation, with no problem between brass and piano. But watch out for that bass drum (kick) rolling down the studio wall.

One other far-reaching problem we should talk about here is the leakage between drums and strings. Even with the drums and the strings well screened, we can spend much time using manymicrophones, very carefully getting tight, crisp snare sounds, forward tom-tom sounds, and good cymbal sounds. But 10m away you've got a large string section with six mics whose gain on the mic channels is somewhere in the region of 40dB per channel (KM84). Open up the string tracks on remix and what have you got? Instant destruction of the drum sound! This is another reason for overdubbing strings. But such leakage can be useful on certain types of sessions where one needs the space and the depth in the recording, such as large internally balanced orchestra where if the drums do sound back it doesn't detract from the overall sound. The live drum sound, as many of you realise, is currently fashionable, particularly on many American pop records and so doesn't really present itself a problem with a live recording of this nature.

One other relevant comment about the sound of the drums on this particular session, and my general personal preference, is to avoid placing the drums within too dead an area. Many acoustic devotees might agree with me: and my preference particularly on this session was to place the drums on a wooden rostrum, partially covered with carpet to stop the kit moving around, which helps the sound enormously. It gives good control but gets a real forward, living sound. Deadening the drummer is the easiest way to destroy one of the hardest working members in the band and to make it sound as if the drummer is playing on cardboard boxes. Another point worth making is that a good room with good acoustic transient response, flexibility in its live and dead areas, right sort of acoustic treatment and correct bass absorption can be made to work very effectively for you, with very little problem hanging over at the console end.

Another point worthy of note is the scored positions of the various instruments. These are notated in **Figure 1**. You'll notice that trumpets 1 and 3 are sitting together, miked separately, with 2 and 4 miked together. There is, though, no hard and fast rule about this. For example, bass trombone always sits on the outside of the trombone section. Now, to get a well-balanced section, particularly when they're playing *divisi*, I prefer to place them as shown. Some engineers may have other preferences and once again no hard and fast rule holds good, so you could have a situation where 1 and 2 would sit together and 3 and 4 would sit together. We could have another situation where trombones 1, 2 and 3 play on one mic and the bass trombone is miked separately or 1 and 2 together with 3 and 4 miked separately. Each of the trombones could be miked separately, but this is getting into an enormous number of channels, particularly with this type of large set-up. It's also interesting to note the horns' seating preferences. 1 and 3 usually sit together. As do 2 and 4. Also, note the position of the bells in **figure 1**, which shows I would prefer to rotate the players to get the bells pretty close together, so that we can get good pick-up and good relation between the written parts. It doesn't normally upset the players when they perform as they might expect to be put in this state unless you have four mics on the horns. And most horn players are very sympathetic to the recording engineer's needs, provided they are not cramped, the recording engineer doesn't get heavy and he understands the musical problems involved.

Stereo

It will be appreciated that the discussion so far has been concerned with multi-track techniques. We have not discussed track layouts and I do not intend to. Any stereo information on remix would be injected stereo information and stereo miking techniques would only be of theoretical mathematical interest on a session of this nature. We could, on this one, handle a stereo string section, but in order to do the job properly it would be my preference to overdub the strings as a pure stereo section. With the type of writing employed, it would be very doubtful whether the right results would be achieved. Further, it would be my recommendation for psuedo stereo. The four mics on the strings would be injected and panned into a stereo position; likewise the mics on the violas and on the celli. We could use stereo techniques on the vocal group, as shown in plate 13. We could use four microphones, cardioids placed in pairs one above the other, each microphone at an angle of 90° to the other – this would give us a stereo picture. Figure-of-eight is avoided because of the likelihood of phase cancellation on the back end of the mic. Also, we don't want any extra room ambience from the other side of the mic, although it's sometimes used to good advantage when the situation demands it. We don't put a single crossed figure-of-eight in the middle because for this type of session it would probably not produce the desired results until the amount of control of balance between the girls and boys.

Conclusion

The really first-class engineer must have the ability, with good basic microphone techniques, to translate the dots from the players and their instruments into a recognisable, fundamental sound that has sympathy and sensitivity for the score, the composer, the musicians, the producer and all concerned with the production. Whatever it is. Therefore the concern in this paper is only to deal with microphone techniques acoustically and not to attempt to deal with equalisation treatments or sound perspectives electronically created through the console. It is appreciated however that the two work closely together and one is used to enhance the other more often than not, It is also extremely necessary that the engineer should understand the generated wave patterns of the sounds they make together with the unwanted odd sounds that they make (with all that plumbing, nobody is perfect). The engineer should understand the generated wave patterns of the instruments and the registers of the instruments that sound naturally good or bad. The comfort of the player must also be maintained when considering microphone techniques – remember, a comfortable player is a happy one and will always give a better result.

In conclusion, I must emphasise once again that mic placement is a very personalised technique. There are so many factors to consider. You need to have the sympathy and above all the sensitivity to interpret what the composer has written, be it group or solo artist, and to work in harmony with the players to create a genial atmosphere for the session – that makes life a lot easier for the engineer on the console. And the mic placements really should come intuitively rather than from a hard and fast written-down rule. Much of this article has been concerned with areas of doubt, which have to be mentioned in order to emphasise this personalised and non-verbal quality. The technique of mic placement remains something *felt* rather than prescribed.

These are but a few idiosyncrasies that occurred within my time at KPM during which we recorded hundreds of tracks, mainly in Cologne, that have stood the test of time. These so-called archive recordings are much sought after. Robin, above all else, was a warm-hearted and caring person; a real person. He worked hard at the career he loved; he supported with passion the composers and musicians that were his lifeblood and expected no less of the people that worked with him. I feel honoured to have spent seven years working with him in mainland Europe, and later at Lansdowne until his retirement, and I know that the music business is a shallower place with his passing in May 2006.

Chapter 14
Building Lansdowne's Reputation

Work continued to flood in so much so that the studio was working almost 24/7 including Preston's work, Denis was cooperative when Bookings asked him to move dates so we could take on clients who wished to work for several days on an album project. Denis held an amount of studio time and he realised we needed to get the income rather than have empty studio days. Anyway, we needed to keep Stevens happy financially. Joe's old perspex screens were falling apart – we dumped them. I asked our builder Alf to make new screens with a window set in (eight and four feet in height) to my specification and covered in heavy cloth to match the studio decor.

I was more concerned about lack of channels on the console; sessions were expanding with larger bands of musos – especially string sections – and we were building a good reputation for our string sounds. The answer to the question of freeing up channel space was to parallel up pairs of string mics using a Y lead, two mic outputs connected together to one mic input and make up the 6dB gain loss on the console. A short-term solution. I spoke with Peter Hitchcock and it was decided we needed a new console built in house by Peter. As one can imagine it went down like a lead balloon with Stevens! Cost, cost, cost! What cost? Well, we use as much hardware as we can from the current EMI console and add another eight extra input channels and extra hardware, mic amps, new faders and some much-needed outboard processing equipment. Denis and I prevailed, Peter would build the console as far as he could in Studio 2 and then we install the old EMI channels in the new desk only keeping the four-track capability.

The extra Fairchild channels contained their own mic amps; we used Fairchild faders, which turned out not to be very accurate. The channel fader gain was controlled remotely by a plug-in light cell – light dependent resistors – in an octal base metal case called Lumiten. Long before VCAs (Voltage Controlled Amplifiers), this varied the voltage from the fader to the lamps in the cell, which shone onto the LDRs, the resistance changed thereby controlling the level. Some faders were OK but we weren't able to cure a level jumping problem permanently at the slightest touch of the fader. The intricacy underneath the fader front panel was that there were turns of wire wound around a plastic tube, along which the fader knob would slide the wiper – in effect a wire-wound linear potentiometer. The trouble was the wire fader slider not making good electrical contact with the wire-wound centre, causing the audio to increase or decrease by up to three dBs when moved – a pain to work with!! Fairchild denied any problem with them! I decided that when I had found a better fader they would be junked! They caused me much grief on an album mix session for Denis mixing the Ken Jones orchestra album called *Swinging the Bard* with an "Elizabethan Consort of Viols", comprising Elaine Delmar on vocals and Geoffrey Emmotts as recorder consort. This was an intricate work requiring absolute mix accuracy that was most frustrating to achieve, – Ken and Denis, who were both present at the mix, couldn't understand why It was taking so long late in the evening. After we had finished the date and Denis was alone I explained my dilemma and established with Denis that those Fairchilds must be removed! Well, I thought, at last I won't get grief from Stevens on this one!

In the meantime, Hitch maintained them as best he could with frequent cleaning – blow the fag ash out, adjust the wiper pressures and struggle on. We finally installed the console in spring '66.

We used the same system to control input level to the Leak TL25s monitor amplifiers: that was tolerable – just.

For the techies – there were light bulbs in each Lumiten (if you are not familiar with these) and each bulb shone onto a light-dependent photo-resistor (LDR). When the applied voltage to the bulbs was varied the signal level was increased or attenuated depending on the bulb brightness on the LDR, the attenuation range was 0 to infinity. Effectively it was an "L" pad. No phase shift is caused. The circuit was unbalanced and the impedance could be 150Ω or 600Ω the insertion loss was 3dB at 600Ω and 5db at 150Ω. Frequency response? Flat DC to 100 kHz. The lamps were powered with 6.3 volts DC. The advantage of this were smooth (supposed to be) level changes, with no wiper noise as was sometimes exhibited by stud-type attenuators, which typically attenuated in ½ dB steps and required frequent cleaning.

Lumitens were used then in audio applications to attenuate a high audio signal while matching the impedance between the source and load in provide maximum power transfer. However, if the impedance of the source is differed to the impedance of the load, the L-pad attenuator can be made to match either impedance but not both. At the time the Lumiten was developed, there were no voltage-control amplifiers. The slider was also cheap and easy to replace if someone spilled liquids on the console.

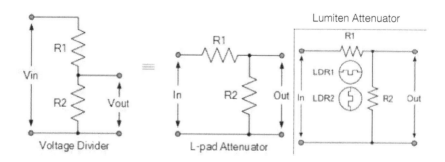

The L-Pad Attenuator is an alternative way of drawing the Voltage Divider

Lansdowne's Busy Schedule

It is an impossible task to speak about every session in detail, as there was so much work; we were turning business away or often clients would simply wait until they could book the studio. Several sessions are worth recalling here.

I was booked to do two sessions on September 30th and November 5th 1964 with producer Norman Petty. Petty was Buddy Holly and the Crickets' producer until Holly's untimely death in a plane crash on February 3rd 1959. Petty also produced Roy Orbison among other artists at his studio in Clovis. His audio recordings were always recorded clean with presence on his tube console.

The arranger was Ivor Raymonde, who brought us much work. It would have made Joe Meek's day to know that Buddy Holly's producer and his recording engineer was working with me at Lansdowne. I was delighted to be doing so. Norman was a musician, engineer, producer and song writer with his own studio in Clovis, New Mexico. An exceptionally prolific producer who had his own trio, the Norman Petty trio with his wife Vi. Petty should never be forgotten for his considerable contribution to the US music industry.

March '65 saw Norman back at Lansdowne with Ivor Raymonde, recording some tracks for Chita Rivera (singer, dancer, actress stage and screen) and some tracks for The Fireballs, New Mexico's first rock and roll recording band.

During 1964, a group called the Barron Knights were signed by Denis. Their recordings engineered by Dave Heelis and between 64 to 68 they had six chart hits. I think they appealed to Denis's dry sense of humour because they parodied the groups of the day including as one example Dave Clark Five whose "Bits and Pieces was parodied as "Boots and Blisters", Freddie and the Dreamers and

Batchelors: and they got around copyright restrictions. The first of their six hits was "Call up the Groups"! The single peaked at #3 in the charts 18 July 64. Denis never ceased to amaze us with new ideas of recording concepts and taking different musical paths, after all was known as a jazz producer. He never moved ahead without meeting with me or other engineers to work out how we should proceed with the recordings.

There was a very fine Canadian violinist named Steven Staryk, a virtuoso who was the concertmaster of Concertgebouw, the Amsterdam Chamber Orchestra. DP wanted to record Steven. He met with him and they decided Steven would record Paganini's *24 Caprices* in the form of *Études for solo* violin, difficult pieces to play. To record Steven – this was 1963 – it was decided we would go to St Gabriel's church in Cricklewood and use the Pye mobile (remote) recording unit with technical engineer Ken Atwood – a man with a wonderful sense of humour who knew his stuff, always had a fag (cigarette) hanging from his mouth. Unfortunately, although the church had good acoustics, the low frequency traffic noise was not acceptable so the session was abandoned. I was asked to recommend another location, this time a "quiet one free of traffic noise!" The other place I knew was St Mary's church, Harrow-on-the-Hill, near Harrow School. We got permission from the vicar, no doubt in return for a contribution to the church fund! The problem was that access was limited and we were in a transit van. Ken said there was nothing for it but to drive the van across the graveyard. There were deep ruts in the grass as it was wet: the vicar didn't say anything – our contribution must have been considerable!

Once installed, we set up and started recording. After the first playback Steven turned to me and said, "You make me sound like a tin fiddle!"

Oops! It wasn't true but he wasn't happy that the pieces were so difficult to play. Steven was playing on a Stradivarius he had bought for £10,000 (today almost impossible to put a price on such an instrument but it would be over a million pounds plus) and he wanted it to sound exactly as he heard it while playing. We made some mic-placing adjustments and a different choice of mic (no eq used) finally settling on the KM54 (tube mic), Steven relaxed and we got on with the recording. We completed the recordings and back at the studio I made a number of very difficult edits and we had a good album.

I recall Denis called me in to a meeting as he frequently did and he told me he had an idea for future recordings. He was going to create the Lansdowne String Quartet and he shrewdly put together a quartet comprising first violin, Jack Rothstein, second violin, Antony Gilbert, viola, Kenneth Essex and cello, Charles Tunnell. I recall they practised for hours in Denis' large office as they were not used to playing together as a quartet – they were some of the "A" listed session men. We discussed how I was going to approach the recording – hitherto string quartets have always been recorded in a large acoustic space with two microphones, for stereo and sometimes ambience mics as well. My approach was entirely different in that I used one microphone (KM54 tube) over each instrument about a metre above the violins and viola and U67 for cello and to create a more open sound used artificial reverb using the EMT stereo plate. The classical aficionados, Allen Stagg being one, said you can't close mike a quartet and record in a pop studio! Old classical dyed-in-the-wool mind-set thinking! They said it wouldn't work but in fact it worked very well. The quartet was most happy with the results.

The first recording was *Sibelius Quartet in D minor*, OP. 56, *Voces Intimae*. That said it all, an intimate recording. It was also successful – the guys appreciated the approach to the way they were recorded. We then went on to record Schubert's *Quintet in C Major OP.163* with Amaryllis Fleming on second cello. The guys and myself were again content with the results – they all played exceedingly well.

We went on to record *"Tony Coe and the Lansdowne String Quartet – Tony's Basement"*, Columbia SCX 6170, this time with the inclusion of bass, percussion (bongos), drums, French horn, vibes and tenor saxophone. Personnel: Alto Saxophone, Tenor Saxophone, Flute, Clarinet – Tony Coe: Tenor Saxophone – Tommy Whittle: Violin – Jack Rothstein, Anthony Gilbert, Viola – Ken Essex, Cello – Charles Tunnell: French Horn – Moe Miller: Piano, Vibraphone – Bill Le Sage: Bass – Dennis Bowden: Drums – Barry Morgan: Bongos – Monty Babson. An especially fascinating combination. David Mack wrote the arrangements. These were very prolific times with Denis and his unusual ideas for recording. Until this point he had been persistently labelled a jazz producer only.

Another of his ideas was East-Meets-West Indo-Jazz fusions and was planning the *Indo Jazz Suite* featuring John Mayer and Joe Harriott (all works composed by John Meyer) to discuss how we go about recording it and the studio set up for musicians it was important for them to have good eye contact and keep good instrument separation without the recording becoming too clinical. It is normal for the Indian musicians to sit on the floor – and no headphones, yes!!

Joe Harriott-John Mayer Double Quintet Personnel: Joe Harriott (alto saxophone), John Mayer (violin, harpsichord), Shake Keane (trumpet, flugelhorn), Chris Taylor (flute), Pat Smythe (piano), Diwan Motihar (sitar), Chandrahas Paigankar (tambura), Coleridge Goode (bass), Alan Ganley (drums) and Keshav Sathe (tabla).

We went on to record another unusual album, this time with compositions by John Mayer and Benjamin Frankel: *Shanta Quintet For Sitar and Strings/ String Quartet No 5* by the Lansdowne String Quartet with Diwan Motihar.

Lansdowne String Quartet Personnel: Jack Rothstein (1st Violin), Tony (Antony) Gilbert (2nd violin), Kenneth Essex (Viola), Charles Tunnell (Cello) and Diwan Motihar (sitar). It was most enjoyable to record and to work in a professional production atmosphere, such a change from pop recording when a producer "used to be certain now wasn't so sure" – about what was required. Believe it or not that was the way it was with some inexperienced producers in the 60s.

I remember in particular another excellent album was with Joe Harriett, the *Double Quintet Indo-Jazz Suite* with a fusion of jazz and Indian instruments. It was not an easy album to record as the Indian instruments' output level was relatively small against the saxophone and the trumpet.

Denis was unique in his "experiments" with musical genres – which was unusual at that time as no one else was doing it or thought about it.

John was a charming man from an Anglo-Indian family, a most talented composer/violinist for whom I had much respect – we got on well together and he had the sensitivity to know what Denis required. I understood at that time that John was a violinist with the Royal Philharmonic Orchestra. Preston went on to work with Mayer until the late 1970s.

In '64 it was becoming a chore for one of us (no assistants) to make tea for the musos and to set up and strip down the studio especially after late night sessions. We had to help musos hump their heavy instruments, tune percussion, guitar amps and Hammond B3 organs – Alan Hawkshaw's was split into two halves, the Leslie speaker the third part. We needed a studio attendant to help them hump in the heavier instruments, set up and strip down the studio, make tea and generally attend to some of our clients' needs and keep the place clean and tidy. In those days the BBC employed many of the London session players in their television studios at Broadcasting House Langham Place and Aeolian Hall Bond Street for the many shows with music the Beeb produced. All these places had studio attendants and the players expected the same service from us at the independent studios. After many interviews I found the answer: John Pemberton who joined the company in December '64. What a character he was, hailing from St Kitts, he also had a charming wife. He was respected and loved by all of the musicians and we studio staff – a very loyal man and a hard worker. Just after he started I was on a gig and the client required tea – it duly arrived with the cups turned upside down on the tray! You see, he was an ex-British Rail worker and that was the British Rail way.

He had good repartee with the musicians too – one enduring memory was when a session finished at whatever time day or late evening after we told him "no overtime the session's finished" he went on the studio floor clapped his hand together three times (always three) and say "Every da body out, da session is finished, don't want no nonsense, abandon de ship! Put da fags out, go home, abandon de ship!" After some sessions you could cut through the air with a knife, so thick and enveloping was the cigarette smoke on the studio floor. While on about John, he liked a few drinks and one of his sayings he was going to the post office for something or other: for which read quick pint at the pub. I was on a session one evening and went into the kitchen where I found a pretty sloshed John and a parcel on the side containing a coconut some tubes marked *Brylcream* and a couple of other bottles of clear liquid.

He offered me the bottle.

"Would you like a drink, sir"?

"No thanks," I replied. "John, what the hell have you got there?"

"It am rum, sir!"

It didn't look like rum.

"What?!"

"Go on smell it sir," he said, again offering me his glass.

You could see the alcohol fumes rising from the liquid.

"What's the alcohol proof?"

" I t'ink it about 100%, sir!"

"For Christ's sake, John, how did you get this stuff?"

"It came in da post," John smiled.

"You were lucky,"

Presumably Customs and Excise hadn't suspected it was rum.

"Yes sir, it was disguised in the coconut and the other bottles."

"John, I think you better go home to sober up and take that stuff with you."

"I am fine, sir" he insisted.

"No, John, you are not fine. You are pissed. Bloody well go home, OK?"

On another occasion, he was walking home after a late evening session pushing his Raleigh bicycle, having had a few tipples in a bit of a (drinking) session. He was stopped by our local beat bobby (oh where are they now?) taken to the police station and charged: viz being drunk while in charge of one green Raleigh bicycle. The next morning he went before the local magistrate and was fined five bob (5 shillings). I paid the fine, retrieved him and we went back to the studio.

"John," I said, "Don't do that again."

"No, sir."

"And try to stay sober will you?" I asked, to which he gave the usual placatory reply,

"Yes, sir."

It was like talking to a brick wall, he did – sometimes – stay sober because he knew his job was on the line. We could always trust John not to let us down though; he knew when not to imbibe. John was an endearing character and loved by all. On another occasion, engineer John Mackswith had a complicated set-up, which he did the night before his session. John Pemberton had gone home when John Mack arrived back the following morning hoping to have a relaxed session – disaster, John had dutifully stripped down the set-up and stored the mics away. John P was not flavour of the month, the expletives were something to behold!

In 1966, the studio continued to flourish all year and while recording in Cologne with Robin I found the answer to the Fairchild fader problem. Danner logarithmic long-stroke carbon faders were fitted on the Siemens console – beautiful to work with, very accurate with the expanded marked dB linear scale from one third of the way up the fader length. Fortunately they were the same physical size and width as the Fairchilds. Now to get it past LGS! No mean feat because I had to import them direct from the manufacturer. Surprise, surprise, technical explanations were listened to with DP's support; I ordered 20 of them plus four spares. They arrived pretty quickly and Hitch installed them. Oh what joy! No more jumping audio caused by the unreliable LDR Fairchild faders.

During these hugely busy periods we were working seven days a week. Not only were we recording but doing maintenance as well lining up tape machines, changing valves (tubes) that had packed up, including the AC701Ks in M49s, 54s and SM2s and more frequently the EL34 output valves in the TL25, plus amps that packed up through being driven so hard and ordering spare components for maintenance because they didn't last very long – some valves had a short life when driven hard!

I wanted to achieve a better professional studio all round, have a maintenance engineer to take the pressure off Hitch and myself and the others, to change direction and have DP's work on a proper, paying basis albeit heavily discounted. That is to say not just as a jazz studio but as an all-round broad-based studio not just, as some said, just a jazz recording studio. I had to put that myth to bed and expand our client-base further and avoid that jazz recording "handle" which was attributed to us by some. Because of DP and his foresight we recorded many prestigious jazz artists of the day – some of the albums are much sought-after and expensive collectors' items today. As DP got busier and busier a guy called Monty Babson (and a singer) joined Denis to take some of the production work off

DP's back. Monty was quite a character, he had his own secretary, Pat Church, and together they occupied a spare room upstairs in Flat 2 next to DP and Stevens.

On the office side, the room in the downstairs entrance (down a narrow iron staircase and a nightmare to get percussion instruments down) was occupied by Preston, Stevens and Preston's secretary, bookings, switchboard *et al*. I persuaded Denis to move to Flat 1 (located by the main entrance to Lansdowne House) and Lionel to move to Flat 2 and have the large office on that flat's first floor with secretaries on the ground floor. The Accounts office being the ground floor was more convenient for Stevens. Denis was pragmatic. It then freed up space for me to take over their old office and make a proper Reception area and have Bookings in my office. In 1969 that whole area was to be converted into the new control room from a stripped-out shell – no compromises.

In early January 1966, the assistant to the engineers in charge of client bookings, Maureen Baker, gave one month's notice to leave the company – somewhat to my relief. She seemed to be spending more time with Stevens in his office, at his behest, consuming quantities of whisky. A consequence of which was time off work recovering from kidney problems. Her being absent put pressure on the other guys who knew how the Bookings book worked and I had enough to do with the demands of recording sessions. Dave Heelis filled the gap or called clients back. Something had to be done about the situation, and fast.

We placed the matter in the hands of a job agency, Reeds in Marble Arch, but this just unleashed a stream of women who saw a recording studio as fertile ground for fawning over artistes or somewhere to collect their autographs. I wasn't looking for a groupie; I needed someone with secretarial experience who was personable. Interviewing all these young women became wearisome as they weren't interested in the job – just the perceived fringe benefits. After a couple of weeks of futile interviews, I was beginning to despair. One afternoon I had another interviewee booked. I didn't rush to be on time as I was in the middle of recording and couldn't leave the client and besides, by this point I was fed up with timewasters. Three-quarters of an hour after the appointed time I met the applicant and apologised for keeping her waiting – she stood out from the other applicants, not least because she was very smartly dressed in a yellow coat. I complemented her on that coat.

"I bought it in New York," she replied.

Miss Edmunds was very gracious but did point out she had been about to leave as she had given up on my appearing. I asked her about her background and she told me she had been a nanny in an upmarket suburb of Riverdale, New York City. She did have office experience but her shorthand was rusty. She came across as very personable and a classily-dressed lady; someone who could organise things. I explained about the Bookings book, the studio engineering side and that she would initially be mentored by Dave Heelis, who was a very experienced engineer and willing to help.

We had a general chat and she told me she'd come back to the UK on the Queen Mary, as she didn't want to fly. I was very impressed with her so at the end of the interview I told her the salary was £13 a week and one month's trial and, if satisfactory, a pay rise after one month.

"You've talked yourself into a job young lady," I said.

To which she replied,

"That's fine but at the end of the month it will be £14 a week?"

"OK." I replied.

What else could I say!

Miss Edmunds started on Valentine's Day 1966 and soon her salary was rising as fast as the bookings! Although I sensed neither of us liked each other much she was very efficient at her job and well organised

– she was good for the studio and popular with clients. She organised the bookings in such a manner that we were eventually working 24/7.

Just after she started, Miss Edmunds received a letter from a Frenchman called Jean Bouchety enquiring about recording. He wanted to book the studio for a French artist, Michel Fugain, who was apparently successful in France but had heard about Lansdowne and its successes. He requested me to engineer the session. It was the start of a long and successful recording partnership stretching over two decades. Jean and I also became very good friends. He was a foremost French composer/arranger. He told me he had been to Pye Studios but didn't like the attitude of the staff or the sound. Michel was easy to work with and so was his singing group Le Big Bazar. One thing that particularly struck me about Bouchety was his string writing – beautiful scoring and was *au fait* about arranging for studio recording.

Hitch left the studio in August 1966 after giving us one month's notice to return to New Zealand to marry. My plan was that we needed to get a much better, commercially made console for the new control room as the industry was changing rapidly and the old modified and then rebuilt EMI console was no longer suitable for our requirements. Hitch had to be replaced by another engineer, not technical, because of the pressure of work.

Engineer Terry Brown joined the company in July '66 from the old Olympic Studios. Although he had worked with Keith Grant he wasn't that experienced. One of the first jobs I gave him was to leader up a Dave Clark Five album. We nicknamed him Terry 2KC brown because every time he recorded it was heavily EQ'd at 2K and it cut your ears off! That's the only frequency we thought he knew, although he did some excellent work. He recorded the *Homburg* single for Procul Harum, *Mellow Mellow* for Donovan, which, with Mickie Most producing was released in February '67 and peaked in the charts at #8. He also worked on Bonzo's first album for Gerry

Bron and Bookings assigned him to work with Monty Babson on jazz albums for Denis Preston. Despite his nickname, Terry was an excellent engineer. He resigned in September '67 to return to the new Olympic now located in Barnes (Middlesex). At that time Monty Babson who produced for DP partnered up with drummer/ percussionist Barry Morgan, Leon Calvert, session trumpet player, Monty Babson - singer and Jerry Allen the Hammond organist to build a new studio: Morgan Studios. Terry had got to know Monty and Barry when working at Lansdowne, they worked together on many of DP's albums.

They approached him to help put Morgan together. They chose CADAC – it was the first commercial console, outside of Lansdowne, produced by the company. After a spell at Morgan, Terry eventually emigrated to Canada. He went on to become a most successful recording engineer/producer in Canada.

One particularly successful artist from France was Eric Charden. Bouchety wrote the arrangements for a single called "Le Monde est Gris, Le monde est Bleu". Technically it was interesting to record because Bouchety had written drum breaks, Kenny Clare on drums, and he wanted us to make an effect with the breaks. So I expanded the drum signal into four seconds plate reverberation and gated it on the return. When the single was released in 1968 it was a massive European hit that summer – I even heard it on the beach over honky Tannoy-type beach loud speakers while on holiday in Rimini, Italy! The drum effect was years before Hugh Padgham, engineer producer (ex Lansdowne engineer from the '70s) created the gated reverb drum sound on Phil Collins' Genesis work. Having had the hit in Europe, I thought we'd see Charden again – it was not to be. None of us could fathom that out as his single was a massive hit that summer so surely he would return. However he was a successful artist in France. Jean Bouchety worked with a Spanish producer Alain Milhaud, from Madrid, and introduced him to Lansdowne. We did much work for Alain right into the 70s. One endearing memory was recording Los Bravos – a film score written by Bouchety for a film called Bravos II. Having spent some days recording the score and over dubbing we went on to mix down the tracks. It got to 11.00pm this particular evening and we had worked continually for 13 hours with breaks on the job! The mix was finally completed, there were some edits to make on the finished master which I suggested we do later. After the mix completion Alain said, "Now we do the film mix!" I wasn't so sure. "Alain surely this album mix should suffice "we haven't recorded the album to picture"! We have no pictures to mix to and my ears are shot after all these concentrated hours. We should come back after a rest to mix the rest with fresh ears" – in my opinion a total waste of time but good income for the studio. It was then I thought how wonderful it would be if we could put the mix in a "black box" to give us manipulation of the mix. A quarter of a century later it happened, recording to and mixing from hard drive. Alain was a nervous kind of guy who it appeared lived on his nerves and chain smoked, the strongest cigarette smoke I ever smelt, with Bouchety smoking American Pall Mall, however he was courteous and for a producer listened to advice! I did the film mix but with hardly any difference in the mix – hey ho that's producers for you. Alain was happy.

When at home one day in the summer of 1967, I was listening to the *Light Programme* (this was before Radios 1, 2, 3, and 4) there was a song being played called, *Let's Go to San Francisco* by the Flower Pot Men, which was a hit (peaked at #4 in the charts) and I loved the sound. I thought, "Wow! That's a good sound, who's the engineer? I must find out!"

A few weeks later, I had a letter from a chap called John Mackswith looking for a job. At the time he was working for Southern Music as an engineer, having left Advision Studios. John takes up the story:

"My friend Gerald (engineer friend at Advision) rang me and said listen there is a job going at Lansdowne, Terry Brown is leaving, he is going to Canada."

John again.

"I thought to myself no way do I stand a chance at getting a job at Lansdowne," he wrote, "and the very next session that came in (to Southern Music) was the rhythm section of Clem Cattini (session drummer who regularly played on many of Meek's productions). And Clem said to me 'John there is a job at Lansdowne – you have to go for it'. I thought 'Oh no', and then I received another prompt from Gerald and from Clem. They said, 'Haven't you applied yet?' Gerald was angry; Clem swore at me and that resulted in my writing a long and detailed letter to AK mentioning the fact that by chance I was the engineer responsible for the current UK No 1 record. A rather cocky letter perhaps but, I thought, why not – here goes!"

When I interviewed John it turned out he had been responsible for balance engineering *Let's Go to San Francisco*, a recording by whose sound I'd been impressed. On a note of interest this single was cut by session men and singers John Carter, Ken Lewis, Tony Burrows and Robin Shaw. It went into the charts in September 1967 and peaked at #4. Mackswith recorded their follow up at Lansdowne, which was a minor hit.

John joined us on a Monday in September 1967 at a basic salary of £18 a week plus overtime. Southern Music offered him more money to stay but he wanted to come to Lansdowne. After the interview, I asked him to come over the studio for the weekend to get a feel for everything. He came over and had his first session for a French artist,

Michel Polnareff, complete with a 42-piece orchestra. Mackswith takes up the story.

> *"The knees started knocking and I told Adrian, 'I've never recorded a 42-piece orchestra in my life!' He told me not to worry – it would be fine! He suggested I go see Dave Heelis for advice. Dave helped me set up on the Monday evening and at 6.30pm he said cheerio. So there I was! I had never recorded 42-piece before. I think the most I had done was 15-piece. I was shitting myself – trembling in the console, sweat pouring off me. Polnareff didn't turn up until 8pm. By this time I had time to get the feel of it. He said,*
> *"'OK, run it through.' He listened to the first run-through and told the strings to go home. I couldn't believe it. It's taken me ages to get this sound and you tell them to go home! Next run-through, he listened to the brass and said,*
> *"Brass – go home."*
> *And so it went on…*
> *'Woodwind go home."*
> *I was left with the rhythm section.*
> *'No I don't like it – just leave us with the guitar,' he said.*
> *"So my first 42-piece session turned out to be Polnareff and a guitar. I was so upset! And his manager could see I was distressed by the whole evening and the fact that all he went back to France with was a quarter-inch recording of his guitar and voice. She was apologetic for all the trouble and gave me a £20 note! That was my first session at Lansdowne."*

John stayed with the company for a number of years, an exceptionally inventive and talented engineer.

At (another!) meeting with Denis Preston, he told me that Laurie was going to record the cast album for the *Four Musketeers*. Michael

Pertwee had written the book, Harry Secombe the music and lyrics were by Herbert Kretzmer (affectionately known as Herbie). It was too large an orchestra for Lansdowne so we had to find another recording space. I recommended Olympic Studios with Keith Grant, whom I knew very well, legendary engineer for many of the popular rock acts of the 60s then 70s: to name some, David Bowie, Eric Clapton, The Eagles, The Jimi Hendrix Experience, Led Zeppelin. I was asked to come and record the cast album. I was doubtful as Keith was a very gifted engineer and I didn't want to cause any problems between he and I. Laurie was doubtful. However, Denis and Laurie insisted that I go to the date to keep an eye on things.

The date of the recording was about late 1966 and, whilst there was nothing to do as Keith was recording and his usual good-humoured self, I wandered around the building and came across a guy in an office who it turned out was one of the maintenance engineers (the studio was working 24/7). He told me he was working 12-hour shifts and was pretty fed up with the relentless pace. I told him I was looking for a technical engineer for Lansdowne and at first he seemed reluctant. However, we met later and agreed he could continue his freelance work and would be given an office in Lansdowne to do his work. The gentleman's name was Clive Green and he came to Lansdowne in Feb 1967, on a freelance contract, and had an office on the first floor of our accounts department. Clive was a real gent - always with an immaculate hand tied bow tie - and he had a wonderful demeanour. Unbeknown to me, it was the start of a long working relationship that would culminate with the founding of CADAC in 1968.

The work rolled in, thanks to Miss Mary Edmunds, and I wanted to put Lansdowne on the map as *the* place to record! During the 1960s, and all through my career, I was privileged to work with the best London session players – the "A-listers". Too many to mention by name but I worked with Jimmy Page – although he couldn't read music (at that time) he was much in demand by arrangers for his

309

playing creativity – and ultimately Led Zeppelin, John Paul Jones, John McLaughlin, Big Jim Sullivan, Vic (Bond) Flick, Eric Ford, Alan Parker, Mo Foster and other "front line" players mentioned in earlier chapters, a very small sample. All dynamic and creative musicians.

As our reputation grew, so did the number of regular bookings. One of which was a pop show for teenagers called *"As you like it"*, produced by Mike Mansfield for Southern Television in 1967. The guest artists were requested by the public on a weekly basis. The show featured the likes of the Bee Gees, Petula Clark and Adam Faith. On one occasion, Cilla Black came to the studio with the Small Faces to record the voice backing tracks, to which they then sang live on the show. Dave Heelis was the engineer.

Musicians' Union rules prohibited artists from reusing backing tracks already recorded and used on other television shows. So a track used for a BBC show could not be played for a Southern Television show and so forth, thereby protecting musicians' jobs.

The MU appeared to be ignorant of the rapid technical changes that were taking place in the industry. The recordings were becoming more technically complex and it wasn't always possible to replicate live the sound produced over many hours in a recording studio. The audience wanted the sounds that they were now used to getting on their record players and radios to be replicated on the TV shows; they wanted to experience the same sounds that we were able to achieve in the studios. The problem was that while a title took six or eight hours, sometimes much more (we didn't hang around ruminating!) if it were a group band, to record a single, television studios were not willing to spend that amount of money for something that was done just to satisfy union rules. In any case, their equipment was not that sophisticated compared to the commercial studios at our disposal. Although titles were rerecorded, the musicians were cut back to a minimum during these sessions. Ultimately the ruling led to a loss of work – an unintentional consequences of union rules!

We only had about 45 minutes or less per title to produce a similar "commercial release" sound for Southern Television. Add to this the fact that the artists would substitute their "proper" (backing) tracks for this "MU-compliant" version or if we had recorded the original at Lansdowne we would do the swop with our copy of the original backing track. It's obvious it was all a farcical waste of time anyway. Often there was a union representative at TV shows to ensure compliance. Don't let the union man see the switch! Anyhow their ears wouldn't know any different.

It soon became clear with the volume of work that the old control room needed to be upgraded. I realised the rapid changes emerging in the recording studio industry, with more multi-tracks being developed such as 16-track and then 24-track analogue machines, meant we needed more space and a larger control room, built from the ground up with a new console design. It was agreed by Lionel and Denis but, as ever, Lionel was mindful of the money. Denis moved out of Flat 1 and into Flat 3 (also owned by the studio), leaving Lionel in flat two with the accounts department, freeing up the space to move the studio reception to Flat 1 and use the old offices' vacated space as the new large control room. The government of the day were offering investment grants for new plant and equipment, so this pleased Lionel. We readily took advantage of the financial help on offer. The government insisted that to qualify for the grant, the new console, including all the individual components, and the cost, had to be logged in detail. The company accountants at the time were Trevor Jones & Co. One of the partners was Eric Scott, with whom Lionel Stevens was in close contact and he sent along an 18-year-old articled clerk, Peter Gregory, to retrieve from the purchase invoices and log every single component bought by the company. The Department of Trade and Industry also insisted, under the Industrial Development Act 1966, that authorised officers of their own be allowed to inspect the assets, to verify that the conditions to which investment grant had been paid were being observed. The process was long, complicated and convoluted with only a small percentage (20%) of the total expenditure on the new equipment allowable for payback. The convoluted process was initiated by Lansdowne in April 1967. It took the inspector from The Department of Trade and Industry until October 1972 to complete his assessment! Nothing new here then! Peter Gregory, now FCA-qualified, came back in 1980 as our company Finance Director! There was a stage when the studio was reimbursed £2,987.00 – the money was slowly reimbursed in tranches – then lo and behold, the government changed its mind, said we didn't comply for a grant (having approved it) and demanded its money back. Naturally, this upset Lionel Stevens! Thank goodness for the APRS (then called Association of Professional Recording Studios) came to our rescue with a newsletter containing advice to its members. Case dropped and all claims met. It took until May 1972! to complete! You see Government typically did not understand studio and record industry jargon.

Clive was to modify the old Hitchcock console, transistorise it and upgrade it for recording on the new American Scully 8-track 1″ recorder of which we were about to take delivery. Initially, we had constant technical problems with the transport's tape tension – a real pain on rewinding tape forward or backwards. This was eventually resolved by Scully – it was one of the first of their machines to be delivered to UK. The old monitor amplifiers were changed from the TL25s to solid state Quad 303s, and another smaller Quad for foldback (one foldback circuit only).

However, before Clive had a chance to modify the console, Lansdowne suffered a smouldering fire in the pre refurbished upstairs control room. John Mackswith recalls,

"Prior to moving into the new (upgraded) control room, I was working with a producer called Jimmy Duncan one evening,

with British pop group *Cupid's Inspiration*. David Baker (Assistant Engineer/Tape-Op, who joined in August '68) was helping me. It must be about midnight and JD is sitting to my right hand side and he looks up and says,

"'I can see something coming out of one of those tiles.'

"I thought, 'Yes it's a pretty old control room, it will go away in a minute.'   .

"'Leave it Jimmy we'll just carry on' I said.

"He looked up and said , 'That tile is turning brown, I think I'm leaving now,'

"And he left. We touched the tile and it was red hot, with smoke billowing behind it.

"'That's it,' I thought '999!'

"At 1.30 am, the fire brigade came in and told us that 'all the equipment will have to come out Guv!' So we moved all the Tannoys and the monitors behind the console. We couldn't disconnect it because of the nature of the cabling but with the help of the firemen we lifted the console out so it was in the middle of the room. They covered everything with tarpaulin and proceeded to gut the ceiling with pickaxes. They took away the first layer of ceiling bricks to reveal the studio wiring, which went through the studio to the control room. The firemen said the wires had been shorting for years as the whole of the conduit was glowing red. Dave and I came in and terminated the power on that conduit, so it was no longer live, and the fire was out. By now it was 4am. So there we are, David and me, stood in a gutted control room with smouldering rubble everywhere and the prospect of a session at two that afternoon.

"'What are we going to do?' I thought. 'I can't phone Adrian as there is nothing he can do.'

"'So instead I phoned our local builder, Alf Williams, who had built the original studio.

"'Before you let ripe (sic) at me for phoning you at 4am,' I said to him, "we've had a fire in the control room, it's been gutted and we have a session at 2pm, can you help?'

"'Don't worry, lad, I'll be around,' he replied.

"So he summoned his crew and arrived at Lansdowne at 6am and in his typical fix-it manner said,

"'Right, let's see what we can do." He re-bricked the conduit that had been taken out by the firemen, cleaned out all the rubble with Dave and I helping him, and he then helped David and I put everything back together. His crew did all the humping, and we did the rewiring. By this time it is 10-11am, and Adrian has arrived, so I had to repeat the story to him. When the session started at 2pm no one was any the wiser as to the events of the previous 12 hours. A week later, David Baker and I were quite disgusted as Lionel Stevens included the generous bonus of £5 in our wage packets for stopping the building burning down!"

Richard Harris Sessions

We had a booking from LA to record Richard Harris. I spoke to John. Was he up for it? We have a batch of sessions coming up and they are all-nighters, starting at 11pm and finishing at 5-6 am. Do you want to do them?

"Yes, of course I do."

"These sessions will go on for three weeks to a month – are you OK with that? What we propose is pay you overtime from 6pm plus your normal wage".

It was a good deal for John and meant the studio was working 24/7. The sessions were vocal overdubs only, on music tracks recorded at SSR (Sunset Sound Recorders) in LA.

John takes up the story:

> *"Some nights Richard Harris never showed which is why it was booked in over the period it was. Prior to the sessions, the eight-track tapes arrived from LA and I had to align our Scully machine to US standards – NAB. Anyway the tapes arrived and I still had no idea who was coming from LA (Harris was on a movie in the UK) or what it was. I put the first track on and it was fabulous – this was not MacArthur Park but one of the other tracks off the album. I'm opening up the faders and it is fabulous. I was listening and this chap pulled himself up to the console, and I said,*
>
> *"'Isn't this wonderful. These arrangements are fabulous!'*
> *"'Yes they aren't bad,' he replied.*
> *"'Who did this?' I asked.*
> *"'I did!' came the reply.*
> *"'Ah,' I said, 'did you write it too?'*
> *"'Yeah,' he said.*
> *"'This is not bad; it's bloody good,' I enthused. 'Who are you?'*
> *"'I'm Jimmy Webb!' he said!*
>
> *"And R H only showed for 50% of the sessions. He was booked to come at 11pm. Didn't come in until 2-3am – pissed so he couldn't do anything and he'd go home. The whole album was done by dropping in line by line. It did take that whole month to do 9 tracks. It was dropping in line by line, with Jimmy Webb going in and singing the line on one track, and with him mimicking and dropping it in the next track. Basically he couldn't sing."*

The album of nine tracks was titled *A Tramp Shining* – Harris's debut album, written recorded and produced by Jimmy Webb. The album included *MacArthur Park*. The cut peaked at #4 in the UK charts and was there for twelve weeks.

The Richard Harris recordings reminded me of some sessions I recorded with twins Paul and Barry Ryan, whose mother Marion Ryan was a popular singer in the '50s when she recorded at IBC. The Ryan twins' sessions were booked by impresario Harold Davison, who produced the boys. As an aside, it was Harold who introduced Dave Clark to America. We did a session, with Ivor Raymonde arranging, recording to four-track. When it came to overdubbing the boys' voices it was a long task to obtain a performance. Me being very patient – sometimes, but getting frustrated about the poor performance, sitting for hours long into the evenings, while they tried to get a decent performance! At one stage, I told Harold things were not improving (in fact it was bloody awful) and their voices were sounding tired so we should stop now (it was rapidly going downhill) and book more time. We did and eventually, after many hours of working on various songs – a result. I mixed the tracks – a no brainer! One song, *I Love Her,* went into the charts in May '66 and peaked at #17. During the long hours of the overdubbing - sometimes word-by-word dropping in on the vocal track - I think Harold was concerned we weren't getting anywhere. Marion came to some of the voice sessions. At the end of it, Harold gave me £20 for my trouble, which I took willingly. Lansdowne's billing for the studio time was not inconsiderable. That was the last I saw of them! A relief.

All in all 1969 was a very busy year! I was involved in so many productions. One in particular I remember was recording a concert at the London Palladium with José Feliciano for RCA Victor records on April 21st. (In 1961 I recorded Chris Barber in Concert there, The *Jazz News* Poll Winners' Concert on Good Friday 31st March 1961). The Feliciano concert was *called Alive-Alive-O! José*

Feliciano in Concert at The London Palladium and it played to a full house. It was very enjoyable concert and easy to record to the eight-track using the Pye Mobile (remote). Personnel: Brian Brockelhurst (bass), Paulino Magalheas (drums & percussion), Peter Ahern (Latin percussion toys), (José Feliciano twelve-string guitar & six-string guitar and vocals) what a superb artist and so easy to work with: a true professional. It was a two-record release. The mixing was done in the US by engineer Hank McGill.

In the middle of 1969, the studio was making good money for Preston and Stevens – in fact, I got the impression from Lionel Stevens that they wanted to sell the studios and retire. - In 1968 Denis *had* done a Europe wide deal with Philips giving them an option to buy the RSL catalogue. We were all invited to celebrate that event with dinner at the Royal Garden Hotel. Bayswater Road. The booze flowed in copious amounts. I was told by Denis there was going to be two visitors one evening, potential buyers, to view the studios when we weren't working and after the staff had left for the day, and I was asked if I could hang around. Paul McCartney and John Lennon were coming to view and would I show them around. I showed them around all the property in the basement including Studio Two (the basement flat) the offices, Flat 1, Flat 2, and Flat 3 and garage. They didn't ask or say much and had a few questions, just looked, were polite and not egotistical as I expected. After they left I thought if they buy the studios I shall no doubt be out of a job. It came to nothing with the Beatles. I was not told why, except that they had decided not to proceed. I believe Denis and Lionel were "testing the waters" for the future. The staff learnt about the two Beatles' visit the following morning, which caused quite a stir and much speculation.

Denis had contacts with EMI Abbey Road producers, Norrie Paramor (Norrie was a shareholder in LRSL) and George Martin, through his Record Supervision Label – The Lansdowne Series was released on EMI's Columbia label. Norrie's title was the Recording

Director for Columbia Records. I believe through these contacts was how the visit came about. The group eventually built their own studios (Apple Corps) at No. 3 Savile Row but that's another story.

Sometime later, when it came to building the new control room out of our old reception/office, we once again contacted Alf Williams.

I also contacted Eddie Veale Associates about designing the acoustics for the new control room and I had many meetings with Clive Green over the design of a new console for Lansdowne. The meetings were to discuss the specialisation and equalisation frequencies required and to stress that the console should have wide bandwidth, somewhere in the order of 5Hz to 75Khz. These frequencies were deemed to be the minus 3dBU points. Clive is a hugely talented technical and creative engineer/circuit designer and it was a marvellous design. We decided it was going to be a new generation of recording consoles – the "Rolls Royce" of consoles – Clive Green was a Rolls Royce devotee and owned a late 1930s model. We wanted the console to be aesthetically different from the staid designs of the day and sonically excellent. Between us we decided to stay away from the drab blue or black panel-sprayed consoles with white lettering of the nearest competitor. The decision therefore was to spray the fader panels in gold with blue silk-screen legends. At this stage there was no plan to start a company and market this new generation of consoles, the idea was simply to build a proprietary one for Lansdowne.

As things started to progress, Clive and I were in agreement that there was a good possibility there was a market for this type of console, and that perhaps we should start a business to compete with the main console manufacturer of the day – Rupert Neve.

Clive suggested he design the circuits and the console constructed by a company he knew in Stansted called Audix. They were already designing and building audio amplifiers and other audio electronic equipment. Clive had a friend, David Bott, who was an engineer working for "TVT" and a meeting was organised at the offices of Charlie Billett, the owner of Audix. Present at the meeting were Clive, myself, David Bott and Charlie. Audix were happy to build the console under the supervision of Clive Green. There was also interest from a new studio – Morgan. It was agreed we should form a separate company, but what to name it? The simple answer was to use our Christian names, first letter of each name Clive, Adrian, David and Charles making a palindrome of our names – this was entirely by accident. So CADAC was formed in 1968. I designed the original logo. More of the company history in Volume II!

There is absolutely no doubt in my mind that the 1960s were pioneering years for Lansdowne. In 1969, we were becoming aware of the changes in recording, new multitrack machines with more tracks available, new outboard processing equipment and new clients booking the studios. I looked forward to this new adventure – as Robin would say, "Into the unknown!"

Adrian eventually married Miss Edmunds. They had four children, five grandchildren and 46 wonderful years together.

Acknowledgements

I wish to thank the following for unstintingly and freely giving their time to answer my voluminous questions in interview, over sometimes long but always pleasant lunches, and also by email and telephone:

Dave Clark, Peter Cox, Bob Butterworth, Mike Brown, Roger Darcy, Chris Dibble, Peter Harris, Patrick Heigham, Laurie Johnson, Simon Jones and Georgina Garrett (for sorting out my timelines and advice), Suzanna Jones, (for her patience keeping me on the right track with my grammar and proof reading), Vic Keary, Kathryn Kerr (for her preparation of my Wikipedia entry), Oliver Lomax (of Dutton Vocalion), John Mackswith, Keith Mansfield, Linda Matthews-Denham for her tireless work on the images for this book Dorothy Marshall (Black & White Minstrels), Hugh Padgham, Richard Preston, John Repsch (for giving me access to his unpublished papers on Denis Preston and Joe Meek), the late Allen Stagg, Ken Townsend OBE and Lucinda Woodward (for her knowledge of the music libraries and their discographies). Special thanks to Anthony Waldron for his technical advice and checking my technical writing.

To François Roux and his staff at La Colombe d'Or, Saint-Paul de Vence, South of France, for taking care of me during my weeks spent writing there, in the most conducive surroundings and weather – not forgetting the most excellent food and wine.

Saint-Paul, April 2016

Volume Two

We hope you have enjoyed reading Volume One of *"Tape's Rolling, Take One" – The recording life of Adrian Kerridge,* which has covered Adrian's career up to the end of the 1960s. Volume Two is scheduled for publication next year, and this will cover from the 1970s onwards, including Adrian's involvement in the advent of digital recording and the growth of music-to-picture scoring at Lansdowne and CTS Studios.

Made in the
USA
Middletown, DE